Southern Space Studies

The Southern Space Studies series presents analyses of space trends, market evolutions, policies, strategies and regulations, as well as the related social, economic and political challenges of space-related activities in the Global South, with a particular focus on developing countries in Africa and Latin America. Obtaining inside information from emerging space-faring countries in these regions is pivotal to establish and strengthen efficient and beneficial cooperation mechanisms in the space arena, and to gain a deeper understanding of their rapidly evolving space activities. To this end, the series provides transdisciplinary information for a fruitful development of space activities in relevant countries and cooperation with established space-faring nations. It is, therefore, a reference compilation for space activities in these areas.

The volumes of the series are peer-reviewed.

Jorge Alfredo Ferrer Pérez ·
Annette Froehlich
Editors

Space Fostering Mexican Society

Developing Mexico Through Space

Springer

Editors
Jorge Alfredo Ferrer Pérez
Unidad de Alta Tecnología
Facultad de Ingeniería
Universidad Nacional Autónoma de México
Juriquilla, Querétaro, Mexico

Annette Froehlich
University of Cape Town
Rondebosch, Western Cape, South Africa

ISSN 2523-3718 ISSN 2523-3726 (electronic)
Southern Space Studies
ISBN 978-3-031-92616-7 ISBN 978-3-031-92617-4 (eBook)
https://doi.org/10.1007/978-3-031-92617-4

This Springer imprint is published by the registered company Springer Nature Switzerland AG
The registered company address is: Gewerbestrasse 11, 6330 Cham, Switzerland

Contents

Historical and Legal Context of Space Activities in Mexico

V. B. Ramos-Barba

Abstract

The development of space activities in Mexico has been characterized by a technological and scientific origin, with a transition to commercial activities, and faces challenges due to the lack of a complete national legal framework. Mexico's space journey began in the 1960s, with notable milestones like the National Commission of Outer Space, the launch of its first satellite, Morelos I, and its first astronaut, in 1985. While Mexico ratified key international treaties, such as the Outer Space Treaty of 1967, it has not developed a specific national law to regulate space activities. The creation of the Mexican Space Agency (AEM) in 2010 was a crucial step in formalizing space efforts, but the absence of detailed regulations has limited the agency's ability to fully govern space activities. This situation has led to gaps in addressing issues like private sector involvement, resources utilization and the capacity building ranging from technology, science, law and social space studies. The purpose of this work is to show the linked context of historical space activities milestones with their law and regulations, expose the challenges and provide important points that need to be addressed for the long-term realization of space activities in Mexico.

1 Introduction

To understand the context of Mexican space activities and their associated regulation, it is necessary to recall of the reached milestones and the rules, laws, and regulations that have existed/exist or the lack of these. It is also fundamental

V. B. Ramos-Barba (✉)
Japan Chapter of the MX Network, Institute of Mexicans Abroad of the Mexican Ministry of Foreign Affairs, Mexico City, Mexico
e-mail: valeria.ramosb@gmail.com

J. A. Ferrer Pérez and A. Froehlich (eds.), *Space Fostering Mexican Society*, Southern Space Studies, https://doi.org/10.1007/978-3-031-92617-4_1

to understand that to enable and exploit the full potential of space activities is necessary the existence of national regulation framework.

2 The Early Years

Mexico started its space activities with the development and launch of the SCT-1 and SCT-2 rockets in October of 1959 and 1960, respectively.[1] At the time of both rockets' development and launch, there was no national law or regulation related with the development, operation, launching, safety and security measures required to handle rockets. In fact, the launching location took place in a private property of the head of the Ministry of Communications and Transport ("Secretaria de Comunicaciones y Transportes" or SCT, and today called "Secretaria de Infraestructura, Comunicaciones y Transportes" in Spanish language) of that period, Walter Cross Buchanan.[2]

Although there was no national regulation, the implementation of such activities was made possible by the political will in favour of space research activities of those who oversaw the former SCT. Later in 1962, a tracking station for the Mercury program of NASA started operation in Guaymas, a city located in the cost of the state of Sonora, in the northwest of Mexico.[3] In terms of regulations, on 19 January 1960, the Telecommunications Law (Ley Federal de Radio y Televisión) was published with concession schemes for commercial, government, cultural, experimental, and other activities of similar nature. A concession was established to be effective for a period of 20 years with an option to be validated for another period to the same licensee. These concessions were granted through public bidding.[4]

On 10 August 1962, the National Commission for Outer Space (known as "Comisión Nacional del Espacio Exterior" or CNEE in Spanish language) was created. It was the national entity in charge of research and peaceful utilization of outer space and its organizational structure was under the SCT. The creation of the CNEE was an important milestone for the country as it established formally an organization to conduct outer space research activities in benefit of the population based on United Nations (UN) recommendations.

The CNEE received as a budget in its first year for research, participation in international events and educational extension around 15% of the total surpluses

[1] RED Universitaria del Espacio, UNAM, "Antecedentes: Cohetería". https://www.astroscu.unam.mx/congresos/rue/Antecedentes_Coheteria.html (all websites cited in this publication were last accessed and verified on 20 December 2024).

[2] Carlos Duarte, "SCT-4: Un homenaje a los inicios de la cohetería mexicana", Hacia el Espacio, Mexican Space Agency. https://haciaelespacio.aem.gob.mx/revistadigital/articul.php?interior=1558.

[3] Victor L. Johnson, "El Valle de Guaymas, lugar estratégico para la NASA en los años 60", El Sol de Hermosillo, 2024. https://www.elsoldehermosillo.com.mx/doble-via/nasa-uso-al-valle-de-guaymas-para-enviar-al-hombre-a-la-luna-11515814.html.

[4] Diario Oficial de la Federación (DOF), "Ley Federal de Radio y Television", 19 January 1960, https://www.diputados.gob.mx/LeyesBiblio/abro/lfrt/LFRT_abro.pdf.

coming from profits from the provision of public services such as railways, post offices and others of the SCT.[5] Among the research subjects of CNEE, these consisted of the following: (a) Sounding rockets; (b) Meteorological satellites; (c) Ballon probes; (d) Bioengineering; (e) Research on basic science and technology; (f) Remote sensing; and (g) Space law.

In terms of space law, the CNEE followed the guidelines from the UN and showed special attention to the following points: (a) establish legal limits between airspace and outer space; (b) registration and control of space objects; (c) the creation of an international organization specialized in space law; (d) the utilization, for the benefit of mankind, of the renewable and non-renewable resources of other celestial bodies; (e) human kind relations with other forms of life on other worlds.[6] Although these guidelines enabled the operation of the CNEE, those were a reference only based on international circumstances of the space race between the United States of America and the Soviet Union. The national context and the growth of technology and services based on space infrastructure were not evident for the Mexican government.

It is worth to mention that during the 15 years of operation of the CNEE there were several technological achievements such as the development of the radio beacons for airplane navigation, and probably, the most representative, the development of the Mitl-1, Mitl-2, Tottl, Tonatiuh, and Hute 2-stage families of rockets as presented in Fig. 1.[7] The CNEE disappear by presidential executive order in 1977 and its competences were transferred to the SCT and the "Secretaría de Programación y Presupuesto".[8]

Between 1966 and 1979 the Treaties of Outer Space were established by the Committee on the Peaceful Uses of Outer Space of the United Nations (UNOOSA). Mexico signed the five Treaties as indicated in Table 1.[9]

Even though Mexico signed all the Treaties on Outer Space, the lack of a national space law continued. The loss of the technical capabilities and experience developed under the CNEE, and even with them, the limited technology capabilities of the country, and poor vision to invest long-term in outer space-based resources utilization for Mexico's growing needs, determined that the country could only meet the demands of the Government and society by outsourcing the use of space systems and services to foreign countries.

---

[5] Diego Emiliano Pazarán Baranda, "La Comisión Nacional del Espacio Exterior", México, 1962–1977, TESIUNAM, 2021, p. 29, pp. 51–57. http://132.248.9.195/ptd2021/mayo/0811633/Index.html.

[6] Ibid., pp. 57–59.

[7] Mexican Space Agency (AEM), "¿Sabías que? - Cohetes y Sputnik", Hacia el Espacio magazine, AEM 2017. https://haciaelespacio.aem.gob.mx/revistadigital/articul.php?interior=592.

[8] Diario Oficial de la Federación (DOF), "DECRETO por el que la SECRETARIA de Programación y Presupuesto dictará las medidas conducentes para proceder a la disolución de la Comisión Nacional del Espacio Exterior", 11 March 1977. https://dof.gob.mx/index.php?year=1977&month=3&day=11&print=true?print=true.

[9] United Nations Office for Outer Space Affairs, *Space Law Treaties and Principle.* https://astro.unoosa.org/astro/instruments-treaties-search.

Fig. 1 Rocket models developed by Mexico between 1959 and 1975 under SCT and CNEE

Table 1 Signature by Mexico of the five treaties of outer space

Treaty	Adoption by the general assembly	Effective date	Signature of Mexico	Ratification in Mexico
The outer space treaty	Resolution 2222 (XXI) opened for signature on 27 January 1967	10 October 1967	27 January 1967	31 January 1968
The rescue agreement	Resolution 2345 (XXII) opened for signature on 22 April 1968	3 December 1968	15 July 1968	11 March 1969
The liability convention	Resolution 2777 (XXVI) opened for signature on 29 March 1972	1 September 1972	29 March 1972	8 April 1974
The registration convention	Resolution 3235 (XXIX) opened for signature on 14 January 1975	15 September 1976	19 December 1975	1 March 1977
The moon agreement	Resolution 34/68, opened for signature on 18 December 1979	11 July 1984		11 October 1991

After the small but prominent period of space research activities in Mexico represented by the existence of the CNEE, the main and longest space activity in Mexico is represented by the satellite communications field. The Olympic Games of 1968 hosted by the country triggered the use of satellite communications to broadcast to several countries the major sports event. To realize the Olympics

Fig. 2 Central telecommunications tower (left), Federal telecommunications network (top right), and construction of the Tulancingo Ground Station (bottom right)

satellite broadcast, Mexico joined the International Satellite Communications Organization (INTELSAT) in 1966, an international organization "…created to own and manage a constellation of communications satellite providing international broadcast services".[10] Mexico's adherence to Intelsat organization allowed the implementation of other programs such as the telematic educational program called "Telesecundaria" in 1965, and paved the way to the start of operations of the National Telecommunications Network, which included the Federal Microwave Network and the Tulancingo Ground Station for Space Communications which started operations in 1968 as shown in Fig. 2.[11]

[10] INTELSAT, "Intelsat History", Official website of Intelsat. https://www.intelsat.com/intelsat-history/

[11] El Mirador Secretaría de Infraestructura, Comunicaciones y Transportes, "La Gesta olímpica de de 1968 por Televisión", SICT Mexico offical government site. https://elmirador.sct.gob.mx/cuando-el-futuro-nos-alcanza/la-gesta-olimpica-de-1968-por-television.

3 Satellite Communications: The Base of Mexican Space Activities

Satellite communication represented an opportunity for the country to modernize its infrastructure as the population grew, and with it, the demand for connectivity-based services. Therefore, the use of satellite communications can be considered the second milestone in the Mexican space activities, and particularly the securing of geosynchronous orbital positions (GEO) with the International Telecommunication Union (ITU) for a satellite network called Satmex, which involved several negotiations with Canada and the United States of America.[12]

To provide a framework for the service of satellite communications in Mexico, the Constitutional article 28 was amended to establish satellite communication as a strategic area so that only the Mexican state would provide this service for the country. The amendment to Constitution article 28 made possible the launch and operation of GEO satellites Morelos 1 and Morelos 2 in June and November of 1985, and the construction and inauguration of a satellite operations centre (SOC) in June of 1985 that provided tracking and control of both satellites.[13] Morelos 1 and 2 consisted of two GEO satellites operating in C and Ku bands based on the HS-376 platform from Huges Space and Communications Company operated in orbital positions 113.5° west and 116.6° west. With the launch of the Morelos 2 satellite, a Mexican scientist was selected to fly as part of the crew of the STS-61B mission of the Space Shuttle Atlantis, which was the launching vehicle of the Morelos 2 and other 2 satellites.[14] This launch of Rodolfo Neri Vela, the first Mexican astronaut, was another milestone reached under the space activities and in particular, satellite communications activities conducted and sponsored by the Mexican Government.

As the first GEO satellites were close to complete their operational life span, in 1995 a new constitutional amendment (to the previous constitutional amendment of 1983) was presented to allow private sector the participation in satellite communications in Mexico, indicating that satellite communication would no longer be a strategic area but would become a priority area. The justification stated that the distribution of data and signals via satellite should initially be exclusive to the Mexican state, given its low profitability at the beginning of its use in the country, until a culture of use and growth in demand was established.[15]

---

[12] Clara Luz Álvarez, "Derecho Satelital y del espacio exterior", Universidad Nacional Autónoma de México, Instituto de Investigaciones Jurídicas, Universidad Panamericana, Mexico 2023, first edition, p. 38. https://archivos.juridicas.unam.mx/www/bjv/libros/15/7352/16.pdf.

[13] Ibid., p. 38.

[14] Shuttle Mission Archive, "STS-61B", Kennedy Space Centre—NASA. https://web.archive.org/web/20031231023443/http://www-pao.ksc.nasa.gov/kscpao/shuttle/missions/61-b/mission-61-b.html.

[15] Clara Luz Álvarez, "Derecho Satelital y del espacio exterior", Universidad Nacional Autónoma de México, Instituto de Investigaciones Jurídicas, Universidad Panamericana, Mexico 2023, first edition, pp. 39–40. https://archivos.juridicas.unam.mx/www/bjv/libros/15/7352/16.pdf

The abrogated Federal Telecommunications Law of 1995 allowed both, national and international companies to provide satellite communication services with a 49% limit on foreign investment, while the decentralized agency Telecomunicaciones de México (now called "Financiera del Bienestar") continued to provide fixed satellite services. As part of the privatization process, on 1 August 1997, the National Satellite Communication Regulation (now replaced) were published. Privatization was carried out through a public bidding process in October 1997. In the same year, in June 1997, Satélites Mexicanos S.A. de C.V. (Satmex) was created. 75% of the shares of Satmex were granted to a private consortium (*Telefónica Autrey, S. A. de C. V./Loral Space & Communications, Ltd.*), while the Mexican Government retained the remaining 25%.[16]

Besides the Constitution article 28 amendment of 1995, the publishing of the National Satellite Communication Regulation,[17] provided the rules regarding national Telecommunications related to satellite communications. Its main points included the following:

- Concessions for national satellites.
- Concessions on foreign satellite signals.
- Permissions for the establishment of transmitting terrestrial stations.
- Signatories of international satellite organizations.
- Provisions common to satellite operators and satellite service providers.
- Services through national satellites.
- Services through foreign satellites.
- Services via international satellites.

3.1 The Mexican Institute of Communications

The National Meteorology Commission and the CNEE, created in the 1960s, were administrative units that set the precedent for research on space applications in the country. Another fundamental entity was the Centre for Research and Development in Telecommunications (CIDET in Spanish), included in the organizational structure of the General Directorate of Telecommunications since 1972 and was in operation from 1972 to 1988. The Mexican Communications Institute (IMC in Spanish) was a legally decentralized agency of the Ministry of Communications and Transportation (SCT), created by Presidential Ordinance on 15 April 1987, with the purpose of solving the national problems faced in communications and generate the conditions that allow the evolution of scientific and technological knowledge in this field to promote the development of the country.[18] In 1988

[16] Ibid., p. 41.

[17] Cámara de Diputados, "REGLAMENTO DE COMUNICACIÓN VIA SATELITE", 1997. https://www.diputados.gob.mx/comisiones/comrtc/CRYT/6.htm.

[18] Diario Oficial de la Federación (DOF), "ACUERDO por el que se crea el Instituto Mexicano de Comunicaciones como un órgano desconcentrado dependiente de la SECRETARIA DE

SCT started the operation of the IMC providing with budget for its operation from 1988 to 1994. The CIDET's resources consisting of laboratories, personnel, and infrastructure were transferred to the IMC.[19] Among the main goals of IMC were[20]:

- Conduct research for similar, adapt and develop technologies related to communications required by the sector.
- Carry out estimates and provide advisory services related to the planning, design, construction and operation of the communications infrastructure.
- Promote the increase of national technology components.
- Design and develop components, equipment and/or systems and promote their application.
- Give an opinion on the evolution of public and private communications services.
- Develop and recommend standards and technical specifications.
- Provide calibration, homologation and testing services for equipment and instruments used in the sector.
- Contribute to the formation of high-level human resources in communications and related disciplines.

The IMC provided technical support to the new generation of GEO satellites operated by the Mexican Government called "Solidaridad", manly related to the requirements definition prior to the international public bidding process for the construction, launch and operation of the Solidaridad satellites system. Under the Solidaridad program, IMC also coordinated and provided the funding for other subprojects:

1. Design of a solar cells-based power system for short-life satellites.
2. SATEX (in Spanish "Satélite Experimental" or Experimental Satellite), a scientific and technology exchange network.

COMUNICACIONES Y TRANSPORTES", 1987. Retrieved from: https://www.dof.gob.mx/nota_to_imagen_fs.php?codnota=4649874&fecha=15/04/1987&cod_diario=200153.

[19] José Luis Peralta, "Capital humano del Ifetel, su recurso más valioso", Etcétera magazine, 21 August 2013. https://etcetera.com.mx/revista/capital-humano-de-ifetel-su-recurso-mas-valioso/.

[20] Eugenio Méndez Docurro, "El Instituto Mexicano de Comunicaciones: una opción al futuro", Revista de Administración Pública, Number 79, Instituto de Investigaciones Jurídicas—UNAM, 1991. https://revistas-colaboracion.juridicas.unam.mx/index.php/rev-administracion-publica/issue/view/1152.

Both subprojects had as a common goal, to create the knowhow and acquire experience in the design, development and operation of experimental Low Earth Orbit (LEO) satellites that could contribute to the technology of commercial communication satellites. However, none of the subprojects culminated in the launching of satellites and/or technology into orbit mainly due to financial problems after the economic crisis in Mexico that started in December 1994, when the initial budget for the subprojects was reduced to half of its original value (1% of the Solidaridad satellites was established to be the budget of the associated subprojects).[21]

The IMC's participation in the World Administrative Radio Conference in 1992 (WARC-92) and subsequent meetings sought, among other objectives, international agreement on the optimal, rational and balanced use of the L-band included in the specified capacity of the Solidaridad satellites.[22]

The Solidaridad-I and Solidaridad-II satellites were launched in 1993 and 1994 by the Ariane 4 rocket, respectively, replacing the first generation of GEO satellites (Morelos 1 and 2). The new satellites included besides the C and Ku bands, the use of the L band for national security purposes.[23] It is important to indicate at this point that both, Morelos 1 and 2 systems, and Solidaridad-I, II systems were operated by the same entity of the Mexican Government, Telecommunications of Mexico ("Telecomunicaciones de México" or Telecomm in Spanish) until Telecomm's satellite operations section was privatized into the company called Satmex in June 1997.

Additionally, the IMC actively collaborated in defining a convention on the establishment of the Regional Centre for Space Science and Technology Education for Latin America and the Caribbean (CRECTEALC) hosted by the National Institute of Astrophysics, Optics and Electronics (INAOE in Spanish).[24] The signatures for the establishment of the CRECTEALC centres, one in Mexico and another one in Brazil, were signed by the Mexican Government and Brazil on 11 March 1997, was approved by the Mexican Senate on 29 April 1997 and entered into action on 18 August 1998 in Mexico.[25] The programs of the CRECTEALC centres are related to[26]:

---

[21] José Galán, "Proyecto *Satex 1*, víctima del *error de diciembre* en ciencia y tecnología", La jornada newspaper, 19 December 2004. https://www.jornada.com.mx/2004/12/19/037n1soc.php.

[22] Luis E. Íñiguez, "El Ing. Eugenio Méndez Docurro y el Instituto Mexicano de Comunicaciones", AGENCIAS ESPACIALES, Actividades, proyectos, investigaciones, posibilidades. Información pública internacional. Aerospace blog, 9 September 2015. https://liniguez.wordpress.com/2015/09/09/el-ing-eugenio-mendez-docurro-y-el-instituto-mexicano-de-comunicaciones/.

[23] Clara Luz Álvarez, "Derecho Satelital y del espacio exterior", Universidad Nacional Autónoma de México, Instituto de Investigaciones Jurídicas, Universidad Panamericana, Mexico 2023, first edition, pp. 42–43. https://archivos.juridicas.unam.mx/www/bjv/libros/15/7352/16.pdf.

[24] Ibid.

[25] CRECTEAL Mexico, "¿Quiénes somos?", CRECTEALC History, INAOE. https://www.inaoep.mx/~crectealc-org/historia.html

[26] United Nations of Outer Space Affairs, "Regional Centres: Background, Mandate and Objectives", UNOOSA—UN. https://www.unoosa.org/oosa/en/ourwork/psa/regional-centres/background.html.

- Remote sensing and geographic information systems.
- Meteorological satellite applications.
- Satellite communications and Geoposition systems.
- Space and atmospheric sciences.

IMC was the last organization to conduct R&D activities under SCT in the XX century. "Although the IMC's main function was to develop studies and research to contribute to the creation of technological solutions in communications, its work emphasized that human capital was a strategic resource. Consequently, one of its main objectives was to contribute to the training of high-level human resources in communications and related disciplines, and this had several handles throughout the nine years of the organization's existence."[27]

The lack of a national strategy, policy, long-term goals and the fact that IMC was created by a Presidential Agreement lead to its demise in 1997. It is important to consider that all previous Mexican organizations related to space research and development and its regulation, have an origin under the Presidential Ordinance, which is a legal act that produces legal effects, made them susceptible to be dismissed by new government policies with different vision.

Along with the activities coordinated by the IMC, a microsatellite project was developed at the National Autonomous University of Mexico (UNAM) during the first years of the decade of 1990, sponsored by the Space Research and Development Program (PUIDE in Spanish) of UNAM, which started its operations on 25 January 1990. The UNAMSAT-I and UNAMSAT-B satellites were developed from an agreement signed between the Radio Amateur Satellite Corporation (AMSAT) and UNAM. The project originated from the interest of Mexican radio amateurs, who consolidated the AMSAT-Mexico organization as an affiliate of AMSAT International.[28] AMSAT provided to UNAM with an advanced engineering package of an amateur satellite, which saved development time, at the time it allowed the modification of the design to adapt it to the requirements of the user.[29] A scientific experiment that consisted of Doppler Shift measurements to estimate the velocity of a meteor entering the Earth's atmosphere was selected as the payload of UNAMSAT-I (based on the experience of meteor scatter communication from the amateur community). UNAMSAT-I, weighing 10.7 kg and with a volume of 10,000 cm^3 (around 23 cm per side) was launched on 28 March 1995, however, it exploded together with the rocket carrying it into orbit.[30] The next year, on 5 September 1996, UNAMSAT-B was launched into a 1000 km altitude at 83-degree inclination orbit from the Plesetsk cosmodrome on a COSMO 3 M rocket. Satellite telemetry was received five hours after launch using a mobile ground station

[27] José Luis Peralta, "Capital humano de Ifetel, su recurso más valioso", Etcétera magazine, 21 August 2013. https://etcetera.com.mx/revista/capital-humano-de-ifetel-su-recurso-mas-valioso/.
[28] Guillermo Núñez Jiménez XE1NJ AMSAT-México, "UNAMSAT. OM-30", radio amateur website, p. 1. Retrieved from: https://0201.nccdn.net/1_2/000/000/17f/728/UNAMSAT-B.pdf.
[29] Ibid., p. 1.
[30] Ibid., p. 3.

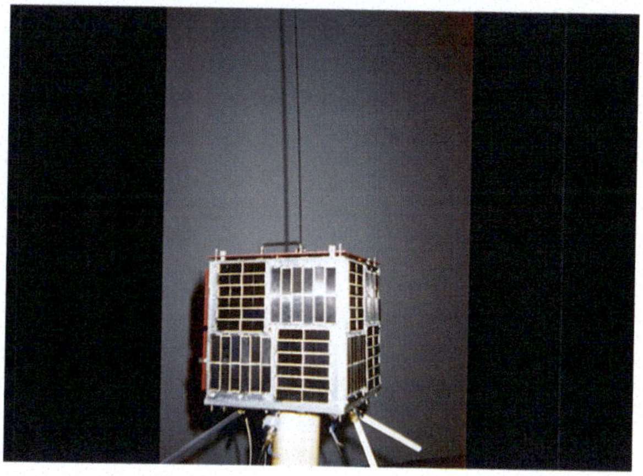

Fig. 3 UNAMSAT-I and UNAMSAT-B Satellites

at 11:00 PM local time (Plesetsk) and in Mexico City at the building of the Instruments Centre where a dedicated ground station was established for the operations. Figure 3 shows the appearance of UNAMSAT-I and UNAMSAT-B.[31]

By the end of the 1990s, with the closing of IMC operations, the privatization of the Mexican satellite operations (Satmex and now under Eutelsat), and with the Mexican Government's vision shifting from space research and development into regulatory and coordination activities mainly, the space activities entered a suspension phase.

4 The Transition from the Mexican State as a Key Sponsor of Satellite Communications Research and Development to Its Regulation

From 1997 to 2010s, there was a lack of governmental entities leading the R&D and coordinating space activities in Mexico. There were a few isolated actions by education and research institutions with limited support from the Government. These institutions mainly consisted of: the National Autonomous University of Mexico (UNAM), the National Polytechnic Institute (IPN), research centres associated to the National Council for Science and Technology (CONACYT, now called CONAHCYT), the Scientific Research and Higher Education Centre of Ensenada (CICESE), the Autonomous University of Nuevo Leon (UANL), the Autonomous University of Baja California (UABC), the University of the Mexican Army and Air Force among others.

[31] Acervo histórico del cómputo en la UNAM, "1993 Satélite UNAM Sat", Galería de fotos, DGTIC UNAM. https://www.historiadelcomputo.unam.mx/unam-sat/.

The joint work of educational institutions such as those mentioned above led to the creation of the Space Science and Technology Network (RedCyTE in Spanish) in 2011, as part of the Thematic Research Networks program of the National Council of Science and Technology CONACYT (now called National Council for Humanities, Science, Technology and Innovation or CONACYHT in Spanish). In addition, the Mexican Society of Aerospace Science and Technology (SOMECYTA in Spanish) was constituted in October 2010, which was registered in the National Registry of Scientific and Technological Institutions and Companies (RENIECYT in Spanish) of the former CONACYT.[32]

RedCyTE was created to bring together all space science and technology efforts at the national level, establish joint projects through collaboration, the realization of meetings, workshops and seminars, as well as the allocation of resources for these activities. It also assisted to the research proposals to be presented to different CONACYT calls and other entities and thus increase the funding options for them".[33] The "Analysis of the Thematic Network of Space Science and Technology (RedCyTE)" of 2021 indicates that "RedCyTE is present in 18 Mexican states and four countries. There are seven fields of knowledge in the Network: Earth and Space Sciences, Technological Sciences, Economic Sciences, Medical Sciences, Physics, Sociology and Political Science. Seventeen Network projects are registered and are developed in several of the 86 affiliated agencies, which are classified as follows: 56 National Institutions, nine Research Centres, ten Companies, six Foreign Institutions, four National Laboratories and one Independent".[34] Among the activities carried out by RedCyTE, some of them intended to have an impact on social and national policies in the health and space areas to name a few.

RedCyTE pushed the generation of Mexican Official Standard (NOM in Spanish) through its participation in the National Technical Council of Space Regulations (COTENE in Spanish), created by the Mexican Space Agency (AEM). Likewise, it participated with the Ministry of Health of the State of Jalisco in the detection of Chagas disease.[35] A new law covering Humanities, Sciences, Technology and Innovation was published on 8 May 2023 which led to the rename of CONACYT into CONACYHT as the new entity in charge of humanities, science,

32 LATAM Satelital, "México: Cooperación Espacial Nacional", Tecnología Espacial LATAM Satelital, 23 August 2016. https://latamsatelital.com/mexico-cooperacion-espacial-nacional/.

33 CUDI, "Redcyte, mexicanos unidos por el espacio", Corporación Universitaria para el Desarrollo de Internet (CUDI), 21 July 2016. https://cudi.edu.mx/noticia/redcyte-mexicanos-unidos-por-el-espacio.

34 Carlos Romo Fuentes, "Diagnóstico de la Red Temática de Ciencia y Tecnología del Espacio (RedCyTE)", Carlos Romo Fuentes on LinkedIn, 2021. https://es.linkedin.com/pulse/diagn%C3%B3stico-de-la-red-tem%C3%A1tica-ciencia-y-tecnolog%C3%ADa-del-romo-fuentes.

35 Ibid.

technology and innovation in Mexico.[36] There is no explicit description about the Thematic Research Networks program in the 2024 Annual Work Program of CONACYHT,[37] and it is unclear the status of RedCyTE at the time of this publication.

The Special Program for Science, Technology and Innovation for the period 2013 to 2018, contemplated the support for the creation and consolidation of national laboratories through the CONACYT.[38] As a result, in the period of 2013 to 2014, the Faculty of Engineering (FI in Spanish) of UNAM received from CONACYT the support to create the National Laboratory for Space and Automotive Engineering (LN-INGEA in Spanish) located at the High Technology Unit (UAT in Spanish) of the FI-UNAM in the Mexican state of Querétaro.[39] LN-INGEA received funding from CONACYT and UNAM for the acquisition of scientific and technological infrastructure, its renewal or replacement, as well as contribute to its consolidation. The laboratory continues its operation and has an active relevance for the space education in Mexico as it is part of the Aerospace Engineering degree created within UNAM in 2020.[40]

Similarly, to the LN-INGEA, another national laboratory supported by CONACYT was created. The National Space Weather Laboratory of the Geophysics Institute of UNAM and the Autonomous University of Nuevo León (UANL) is integrated by the Space Weather Service Mexico (SCiESMEX), the Institutional Space Weather Repository (RICE), the Space Weather Supercomputing Centre (CESCOM) and the Space Weather Instrumentation Network of UNAM and UANL.[41] Besides the support of CONACYT and the mentioned universities in Mexico, the origin of LANCE relies on the recommendations originated from the Committee on Space Research (COSPAR) in 2015 on Space Weather (SW). The same recommendations highlighted the need to strengthen and increase

[36] Diario Oficial de la Federación (DoF), "LEY GENERAL EN MATERIA DE HUMANIDADES, CIENCIAS, TECNOLOGÍAS E INNOVACIÓN", 2023. https://www.diputados.gob.mx/LeyesBiblio/pdf/LGMHCTI.pdf.

[37] CONACYHT, "Programa Anual de Trabajo 2024 Consejo Nacional de Humanidades, Ciencia y Tecnologías", CONACYHT 2024. https://www.siicyt.gob.mx/index.php/transparencia/informes-conacyt/programa-de-trabajo-1/5017-programa-anual-de-trabajo-2024/file.

[38] Plan Nacional de Desarrollo 2013–2018, "Programa Especial de Ciencia, Tecnología e Innovación 2014–2018, Logros 2015", Plan Nacional de Desarrollo 2013–2018. https://www.siicyt.gob.mx/index.php/transparencia/informes-conacyt/logros-programa-especial-de-ciencia-tecnologia-e-innovacion-peciti/1488-logros-peciti-2015/file.

[39] LabUNAM, "Laboratorio de Ingeniería Espacial y Automotríz LINGEA", Scientific Research Coordination UNAM, pp. 14–16. https://labunam.unam.mx/micrositios/LINGEA/antecedentes.php?il=MjYy.

[40] GACETA UNAM, "Crea la UNAM la licenciatura en Ingeniería Aeroespacial", Gaceta UNAM, 12 February 2020. https://www.gaceta.unam.mx/crea-la-unam-la-licenciatura-en-ingenieria-aeroespacial/.

[41] FAQS LANCE, "¿Qué es el LANCE?", National Laboratory of Space Weather. https://www.lance.unam.mx/acerca-de/faqs/.

the investment and development of scientific infrastructure in observation, analysis and prediction of SW. The Space Weather expert group of UNOOSA and the Inter-programme Coordination Team on Space Weather (ICTSW) of the World Meteorological Organization (WMO-UN) also recommended similar actions.[42] On the national framework, the General Law of Civil Protection (LGPC in Spanish) in Mexico was published in 2012 and modified in 2014, 2018 and 2023, including explicit mentions of possible effects in the country due to SW phenomena in its article 20 (included in 2014).[43]

The disappearance of public telecommunications monopolies in several countries around the world has been driven by a combination of regulatory reforms and market liberalization, which have led to the transition from state-controlled monopolies to competitive markets. In Mexico, the government also went from being a service provider to a regulatory entity. From the governmental side, the old IMC was replaced by the Federal Telecommunications Commission (COFETEL in Spanish), which was created by Presidential Agreement on 9 August 1996 and was conceived as a decentralized administrative organization under the SCT, with technical, operational, expenditure and management autonomy, in charge of regulating, promoting and supervising the efficient development and broad social coverage of telecommunications in Mexico, with full autonomy to issue its resolutions.[44]

Most of the IMC technical staff moved to the COFETEL and took an important role to start and maintain service market deregulation activities, renewal of the criteria for certifying equipment, in the verification of the applicable standards, technical support for taking a position in international meetings on standardization, especially with the United States and Canada, within the framework of the North America Free Trade Agreement (NAFTA). However, the COFETEL also had to recruit personnel in other disciplines as its work also demanded additional professional expertise, especially in the disciplines associated with an integrated regulatory structure: law and economics. "The economic contribution was essential to address key issues related to costs and tariffs of services, including investment and expenditure projections, and to assess the financial equilibrium in the operation of the services. The legal aspect is equally crucial to design regulations and support the correct exercise of the provisions and mandates in force, while aiming to provide legal consistency, relevance and ethics of their application, and to consolidate the public interest in the sector."[45] By combining the technical expertise with law and economics, COFETEL was completely established as a regulatory organization.

[42] Ibid.

[43] Diario Oficial de la Federación (DOF), "Ley General de Protección Civil", 6 June 2012. https://www.diputados.gob.mx/LeyesBiblio/pdf/LGPC.pdf.

[44] Diario Oficial de la Federación (DOF), "Decreto por el que se Crea la Comisión Federal de Telecomunicaciones", 1996. https://dof.gob.mx/nota_detalle.php?codigo=4895322&fecha=09/08/1996#gsc.tab=0.

[45] José Luis Peralta, "Capital humano del Ifetel, su recurso más valioso", Etcétera magazine, 21 August 2013. https://etcetera.com.mx/revista/capital-humano-de-ifetel-su-recurso-mas-valioso/.

The legal base to enable the creation of the COFETEL were the Federal Telecommunications Law of 1995 (LFT95), and the last constitutional amendment to article 28 that allowed the participation of the private sector in telecommunications in Mexico. The Mexican government aimed to promote competition in telecommunications, provide legal certainty to investment and train specialized human resources through the attributions granted to the SCT by the LFT95 and being implemented under the COFETEL.

In 2013, there were new constitutional amendments to articles 6, 7, 27, 28, 73, 78 and 94 as part of series of constitutional reforms presented by the Mexican government in office. As a result of the 2013 reforms, the COFETEL was ended and a new autonomous constitutional entity was created, the Federal Telecommunications Institute (IFT in Spanish). The IFT was created on 10 September 2013 directly from the amendments to article 28. The IFT is an autonomous organization which has as its main objective the efficient development of broadcasting and telecommunications in Mexico in accordance with the provisions of the Constitution. To achieve this, it must regulate, promote and supervise the use, development and exploitation of:

- The radiofrequency spectrum.
- The networks.
- The provision of public broadcasting and telecommunications services.

It is also responsible for ensuring equitable access to infrastructure and other essential inputs for information and communication technologies, as well as broadcasting and telecommunications services, including broadband and internet.[46] The IFT supports its activities on two national laws: the Federal Telecommunications and Broadcasting Law (LFTR in Spanish) and the Federal Economic Competition Law (LFCE in Spanish), published on 14 July 2014[47] and on 23 May 2014,[48] respectively. Both, LFTR and LFCE were amended in 2017 and 2021 and remain active to the time of the present publication.

The IFT inherits from COFETEL the status of regulatory entity for telecommunications and broadcasting. On 23 January 2023, the Plenary of the IFT issued the Regulatory Provisions on Satellite Communications, which established the regulatory provisions for satellite communication. The new satellite regulation replaced the *Reglamento de Comunicación vía Satélite* of 1997. Although the Mexican Presidency (Executive Power) filed a constitutional controversy to the Mexican Supreme Court of Justice for considering that the IFT invaded the powers of the Federal Executive and the Congress, the Mexican Court of Justice determined that

---

[46] Instituto Federal de Telecomunicaciones, "¿Qué es el IFT?", IFT website. https://www.ift.org.mx/que-es-el-ift/que-es-el-ift.

[47] Diario Oficial de la Federación (DOF), "Ley Federal de Telecomunicaciones y Radiodifusión", Mexican Congress. https://www.diputados.gob.mx/LeyesBiblio/pdf/LFTR.pdf.

[48] Diario Oficial de la Federación (DOF), "Ley Federal Federal de Competencia Económica", Mexican Congress. https://www.diputados.gob.mx/LeyesBiblio/pdf/LFCE_200521.pdf.

the controversy filed by the Mexican Presidency was admissible but unfounded, considering that "if the law orders another body to regulate its content, then the Executive Power cannot regulate that same provision without contravening the principle of supremacy of the law, since it would be acting against the law itself",[49] therefore, the satellite communication regulation of 2023 remains as the current regulation in the country related to the use of satellite communications at the time of this publication.

Regarding the Mexican Space Policy, the government has mainly focused on the satellite policy giving it priority over the whole space activities. The government published the Federal Government Satellite Policy (PSGF in Spanish) on 15 May 2018 and amended it on 21 December 2023. The PSGF mentioned as its main purpose that "The general objective of the Federal Government's Satellite Policy (PSGF) is to exercise the State's leadership over satellite development in Mexico and its orbital resources, as well as its regulation, considering social inclusion, national security, economic development, technological development, international cooperation and the space ecosystem, strengthening its sovereignty and autonomy in space matters for the benefit of the population.".[50] The PSGF contains six main guiding principles with their respective strategic actions. The implementation considers the coordination through a National Satellite Program, based on the National Development Plan that is executed by the SICT and the AEM, in coordination with the IFT. The six principles of the PSGF consist of[51]:

(1) State governance and sovereignty over national development.
(2) Satellite services for social and digital inclusion.
(3) Satellite services for economic development.
(4) Satellite services for national security.
(5) Technological development and satellite ecosystem.
(6) International cooperation.

The new satellite policy considered the creation of a Satellite Advisory Council (CCS in Spanish) with the purpose of coordinating the implementation of the strategic actions of the PSGF. The CCS would be formed by the public and private sectors, academic and research institutions, whose mission will be to concentrate

[49] Suprema Corte De Justicia de la Nación, "CONTROVERSIA CONSTITUCIONAL PROMOVIDA POR EL PODER EJECUTIVO FEDERAL, DEMANDANDO LA INVALIDEZ DEL ACUERDO MEDIANTE EL CUAL EL PLENO DEL INSTITUTO FEDERAL DE TELECOMUNICACIONES EMITE LAS DISPOSICIONES REGULATORIAS EN MATERIA DE COMUNICACIÓN VÍA SATÉLITE, PUBLICADO EN EL DIARIO OFICIAL DE LA FEDERACIÓN DE VEINTITRÉS DE ENERO DE DOS MIL VEINTITRÉS", Supreme Court of Justice of Mexico, 25 April 2024. https://www.supremacorte.gob.mx/sites/default/files/versiones-taquigraficas/documento/2024-04-25/25%20de%20abril%20de%202024%20-%20Versi%C3%B3n%20definitiva.pdf.
[50] Diario Oficial de la Federación, "ACUERDO que establece la Política en materia satelital del Gobierno Federal", 2023. https://dof.gob.mx/nota_detalle.php?codigo=5712152&fecha=21/12/2023#gsc.tab=0.
[51] Ibid.

the efforts of the various agencies and the industry for the development of the Mexican satellite field. It is important to mention that the PSGF considers satellites in general, although it focuses strongly on satellite communications applications by explicitly mentioning them throughout the document, and making a brief mention in the section on the importance of satellites in Mexico (section 1), the background (section 2), and the principle 4 (Satellite services for national security) to other satellite applications such as geo-localization, Earth observation, navigation and GEO-positioning.

The period of 1997 to the decade of 2010 consolidated the transition of the Mexican government from conducting R&D in the field of satellite communications to its regulation and policy, space research and development has been left to public and private educational and research institutions with the support of the National Council of Humanities, Science, Technology and Innovation. As such, CONACYHT's support has been limited considering the variety of institutions, projects and scopes that Mexico's space development requires. In addition, there is a lack of laws and regulations on all space activities, with efforts focused only on satellite communications applications, and mainly on their regulation. There is also a lack of a long-term vision of the future of satellite applications in general and of their funding, as well as their isolation from other space activities, which compromises the PSGF itself.

5 The Mexican Space Agency and the New Generation of Space Enthusiasts

On 25 November 2005, an initiative for the creation of the Mexican Space Agency was presented in the Chamber of Deputies of the LIX legislature (2003–2006), which was approved in the Chamber of origin and sent to the Senate on 26 April 2006. The Senate's Science and Technology Committee took on the task of incorporating the observations of the Mexican Space Agency's Promotion Group, as well as those issued by CONACYT and the Science Advisory Council of the Presidency of the Republic, into the bill approved by the Chamber of Deputies. The integrated proposal was unanimously approved on 4 November 2008 and was subsequently sent to the Chamber of Deputies to be turned over to the Science and Technology and Budget and Public Account Commissions for a second analysis and round of voting. The Chamber of Deputies of the LXI legislature (2009–2012) approved the initiative on 20 April 2010. The Law creating the AEM was promulgated on 13 July 2010 by the President of the Republic, Felipe Calderón Hinojosa; and was published in the Official Gazette of the Federation on 30 July 2010.[52]

[52] Gobierno de México, Agencia Espacial Mexicana. Antecedentes, 13 July 2011. https://www.gob.mx/aem/acciones-y-programas/antecedentes-de-la-aem#:~:text=La%20Ley%20que%20crea%20la,30%20de%20julio%20de%202010.

The Mexican Space Agency is a decentralized agency sectored to the SICT. Unlike the past government entities created by Presidential Decree to carry out the Mexican government's space-related activities and applications, the AEM was created by law. This means, while a law contains abstract provisions of general observance, a decree is made up of provisions of a particular nature and its validity is limited in space, time, places, corporations, establishments and/or persons. A decree may be issued by either the Executive or Congress.[53] Based on article 2, Sections VIII and XI of the Law that creates the AEM (LAEM), the Agency's purpose is: "to serve as an instrument of the State's leadership in this field, in order to strengthen sovereignty and receive from public, private and social entities, proposals and observations in the space area for their study and consideration".

At present, the article 28 of the Political Constitution of the United Mexican States (CPEM) establishes that satellite communication is a priority area for national development. There is a national constitutional reform proposal for amendments to articles 28 and 73 of the CPEM so that the Congress shall have the power to legislate on outer space activities and that these are considered priority areas for national development under the terms of article 25 of the National Constitution. The transitory articles of the initiative propose to generate the National Law on Space Activities and secondary regulation of the sector.

In October 2020, in a first attempt, Deputy Marivel Solis Barrera of the LXIV Legislature (2018–2021) presented a reform initiative[54] that was not ruled on and was left without processing. The proposal comes from the doctoral thesis of the author Luis Antonio López-Velarde Sandoval.[55] In a second attempt, the Deputy of the LXV Legislature (2021–2024), Jesus Briano Borunda, presented the initiative on 15 March 2022, which has been approved in the Chamber of Deputies on 30 March 2023 and sent to the Senate where it has been approved by committees on 24 April 2024 and presented for plenary opinion on 25 April 2024 currently.[56]

In matters of public policy, Mexico's policy on outer space comprises:

- The General Guidelines of Mexico's Space Policy (LGPEM). These are the outcome of consultation forums ordered in the Third Transitory article of the Law that creates the Mexican Space Agency (LAEM). The guidelines establish nine strategic objectives and 13 General Aspects of the Mexican Space Policy.

---

[53] Cámara de Diputados. Baena, Juventina. ¿Qué es un decreto? 7 Nov 2023. https://comunicacionsocial.diputados.gob.mx/revista/index.php/abc-legislativo/-que-es-un-decreto.

[54] Mexican Congress. National constitutional initiative to reform articles 28 and 73 by Deputy Marivel Solis of the LVXIV Legislature, 14 October 2020. http://sil.gobernacion.gob.mx/Archivos/Documentos/2020/10/asun_4091344_20201014_1602716063.pdf.

[55] Luis Antonio López-Velarde Sandoval, El Espacio Exterior y su Regulación, Ediciones ECOE, 2018.

[56] Secretaria de Gobernación, "Sistema de Información Legislativa". http://sil.gobernacion.gob.mx/Busquedas/Avanzada/ResultadosBusquedaAvanzada.phpSID=&Serial=fd4899f37e6e6abb5f7a99cd6ee56971&Reg=7&Origen=BA&Paginas=100#5.

- The National Space Activities Plan (PNAE). It establishes the policy guidelines, objectives, strategies and lines of action aligned with the National Development Plan (PND) 2019–2024 and the Sector Program of the Ministry of Communications, Transportation, and Infrastructure.
- The Federal Government's Satellite Policy (PSGF). Published on 15 May 2018 and amended it on 21 December 2023. The PSGF contains six main guiding principles with their respective strategic actions. The implementation considers the coordination through a National Satellite Program, based on the National Development Plan that is executed by the SICT and the AEM, in coordination with the IFT.

Regarding Mexican space related standards, with the support of the General Directorate of Standards of the Ministry of Economy (SE-DGN), on 22 March 2016, the Technical Committee for National Space Standardization (COTENNE) was created. It has the purpose of developing Mexican Standards for materials, component and equipment for the development and operation of aircrafts and space vehicles. These standards come from the Federal Metrology and Standardization Law (LFMN), published in 1992 with the intention of standardizing metrology processes, services and laboratory tests. According to the LFMN, there are basically three types of standards: Official Mexican Standards (NOM), Mexican Standards (NMX) and Reference Standards (NRF).

NMX are developed by a national standardization body or the SE-DGN. They establish the minimum quality requirements of products and services to protect and guide consumers. Their application is voluntary, except in those cases where individuals express that their products, processes or services conform to them, or when a NOM requires the observance of a NMX for specific purposes.[57]

On 1 July 2020, a new Law was issued to replace the previous one and its purpose is to establish and develop the bases for industrial policy within the scope of the National Quality Infrastructure System,[58] through the activities of normalization, standardization, accreditation, conformity assessment and metrology. The new law has the purpose to promote economic development and quality in the production of goods and services. It is intended to:

- Expand productive capacity and continuous improvement in value chains.
- Promote international trade and to protect the legitimate objectives of public interest described by the new law.

So far, three Mexican Technical Standards have been published as shown in the Table 2 (at the time of this publication), and there are five drafts of Mexican

[57] Gobierno de México, Secretaría de Economía, Mexican Standards. https://www.gob.mx/se/acciones-y-programas/standards.

[58] Diario Oficial de la Federación, "DECRETO por el que se expide la Ley de Infraestructura de la Calidad y se abroga la Ley Federal sobre Metrología y Normalización", 1 July 2020. https://www.diputados.gob.mx/LeyesBiblio/abro/lfmn/LFMN_abro_01jul20.pdf.

Table 2 The current Mexican Standards (NMX) related to space

Designation	Standard name
NMX-AE-001-SCFI-2018	Space Systems—CubeSat design, requirements and classification
NMX-AE-002-SCFI-2019	Space Systems—Risk management
NMX-AE-003-SCFI-2021	Space Systems—Electromagnetic compatibility requirements

Table 3 Drafts of Mexican Technical Standards

Category	Subcategory
Space systems	Harnesses—Crimping
Space systems	Structural Assemblies and Components (ICS: 49.140)
Space systems	Design, qualification, and acceptance testing of small space vehicles and units (ICS: 49.140)
Space systems	General test requirements for launch vehicles (ICS: 49.140)
Space systems	Structural design—loads and induced environment (ICS: 49.140)

Norms (NMX) being reviewed by the Committee, which are briefly described in the National Quality Infrastructure Program 2023 as shown in the Table 3.[59]

Besides the AEM creation and its start of operation, the Mexican government created the GEO satellites program called MEXSAT (Satélites Mexicanos in Spanish) in 2007 based on the Strategic Plan of Satellite Communications of that year between SCT, the Ministry of National Defence (SEDENA), the Ministry of Navy (SEMAR), the Federal Police (PF), the Attorney General's Office of the Republic (PGR), the National Security Research Centre (CISEN in Spanish), and Telecommunications of Mexico (TELECOMM).[60] In 2010 SCT designates TELECOMM as the operator of the new satellite system, including the administration of two new satellite operations centres (SOC) and two network operations centres (NOC) located in Hermosillo City in the Mexican state of Sonora (northwest Mexico) and in Mexico City. The first satellite called Bicentenario (Mexsat 3) was launched on 19 December 2012 from French Guiana onboard an Ariane 5 rocket and provides fixed satellite services in C extended band and Ku extended band at the orbital position 114.8° West. The Bicentenario satellite is based on the Start-2 BUS of

[59] AEM. Impulso a la estandarización del sector espacial "Revista electrónica Hacia el Espacio", 1 February 2023. https://haciaelespacio.aem.gob.mx/revistadigital/articul.php?interior=1288.

[60] TELECOMM, "MEMORIA DOCUMENTAL, Telecomunicaciones de México, Operador del Sistema Satelital Mexicano Mexsat", TELECOMM. https://finabien.gob.mx/gobmx/wp-content/uploads/2024/04/Memoria%20Documental%20Telecomunicaciones%20de%20M%C3%A9xico,%20Operador%20del%20Sistema%20Satelital%20Mexicano%20Mexsat.pdf.

Orbital Sciences Corporation (now Northrop Grumman) covering the Mexican territory the Gulf of Mexico and the Exclusive Economic Zone of Mexico. A second satellite called Morelos 3 (Mexsat 2) was launched on 2 October 2015 on an Atlas 5 rocket from Cape Canaveral, Florida in the United States of America. It operates from orbital position 113.1° West and provides mobile services in L band and Ku band to the Mexican territory, the Economic Exclusive Zone of Mexico, Central America, the Caribbean and the north part of South America and its surrounding waters.[61] A third satellite called Centenario (Mexsat 1) was launched on 16 May 2015 but after a few minutes from the liftoff, the third stage of the rocket failed re-entering and burning in the Earth's atmosphere with the Centenario satellite onboard. The Morelos 3 and Centenario satellites were part of a redundant satellite system based on the Boeing HP-702 BUS system, with one satellite intended to operate as the backup of the other to reduce the risk of losing operations in L band for national security.[62] With the implementation of the MEXSAT system, the Mexican State recovered its operational capacity for national security communications, disaster response services, social coverage and the safeguarding of the use of orbital positions and frequencies assigned to the country.[63]

5.1 Recurrent Space Events in Mexico

After the AEM started operating and its activities spread to a wider part of the space community in the country, more space activities from education and research institutions, other governmental entities, as well as from the civil society started to bloom. Among the most relevant events include, in chronological order, FAMEX, CONACES and ENMICE; with a special mention of the International Astronautical Congress held in Mexico in 2016.

The Mexico Aerospace Fair (FAMEX in Spanish) is a commercial activity in the areas of civil aeronautics, military, security and defence, whose objective is to bring together the leaders of these sectors to promote commercial exchange and boost the growth of the aerospace industry in Mexico.[64] The first edition of FAMEX started in 2015 and included 240 companies from 16 countries. In the second edition in 2017 the number of companies increased to 505 from 35 countries, the 2019 edition increased the number to 635 companies from 39 countries. Fourth edition reached 317 leading companies from 47 countries and the fifth

61 Ibid.

62 Ibid.

63 TELECOMM, "Telecomm y El Sistema Satelital Mexicano MEXSAT", TELECOMM. https://swfound.org/media/78888/30_telecomm%20y%20el%20sistema%20satelital%20mexicano.pdf.

64 FAMEX, "Feria Aeroespacial México", FAMEX. Retrieved from: https://www.f-airmexico.com.mx/evento.html.

edition in 2023 included 706 companies from 51 countries.[65] FAMEX is organized by the Mexican Armed Forces and although its main composition consists of aeronautical companies and organizations, it has also included the participation of institutions such as AEM, universities and research institutions with space development activities.

The National Congress of Space Activities (CONACES in Spanish) is the national congress of the Mexican Space Agency (AEM) covering a wide variety of subjects related to space science and technology.[66] CONACES includes keynote lectures, workshops, round tables, posters and technical sessions in nine categories[67]:

(1) Space Science, Space Exploration and Astrobiology
(2) Social Sciences, Space Law and Regulation
(3) Earth Sciences and Earth Observation Agriculture, Natural Resources, Early warning, Natural phenomena and disasters
(4) Health Sciences
(5) Education
(6) Launch vehicles, Spacecraft and Materials
(7) Nanosatellites
(8) Space Communications
    a. Geolocation and navigation; GNSS
    b. Optical, satellite, rural communications, etc.
(9) Psychology, Arts and Humanities

CONACES started in 2020 with a remote format, followed by the 2021 editions also in a remote format, and for the editions of 2022 and 2023 with on-site format. The on-site formats allowed the CONACES to be held in different locations across Mexico such as the cities of Aguascalientes (central Mexico), Acapulco (west coast of Mexico) and Matamoros (northeast of Mexico). The participants consist mainly of students, professors and researchers in both, Mexico and overseas, with interest in projects of national and international scope, as well as contributing to the creation and growth of collaborative networks between professionals in the Mexican space sector, and to promote scientific and technological vocations among young people in Mexico and Latin America.[68]

[65] FAMEX, "Welcome to FAMEX", FAMEX. Retrieved from: https://www.f-airmexico.com.mx/en/.
[66] Mexican Space Agency (AEM), "CONACES", CONACES 2024, AEM. https://conaces.aem.gob.mx/.
[67] CONACES-AEM, "Convocatoria CONACES 2024", Mexican Space Agency (AEM). https://conaces.aem.gob.mx/convocatoria#.
[68] Ibid.

The Mexican Experimental Rocketry Engineering Meeting (ENMICE in Spanish) is the leading experimental rocketry event and competition in Mexico, which, with the support of its allies, collaborators and sponsors, is part of the national effort to promote the participation and growth of the Mexican space industry, bringing together, promoting and recognizing national talent in space science and technology. It was founded in 2020 by Ramón Córdova, Felipe Ávila, Ariel Gómez, Walter Ahrens and Jonathan Castillo with the aim of promoting high power experimental rocketry across Mexico.[69] ENMICE brought back the effort to develop rockets in Mexico after more than 60 years of its first attempts to a wide sector of the community interested in rocket science. It has hold competitions of rocket launching in 2021, 2022 and 2023–2024, and has seen success among university students with a growing community in the development and launching of rockets.

Prior to ENMICE there were very few attempts to develop and launch rockets in Mexico. One of these is the case of Luis Arellano Leeuwenberg, a professional working on industrial thermal systems in the city of Saltillo, in the state of Coahuila located in northeast of Mexico. Luis Arellano has developed both, solid fuel and liquid fuel rockets, and launched several of them in more than 20 years.[70] Another effort is represented by Luis Gerardo Saucedo Zárate who founded the organization called "Instituto Mexicano del Espacio Ultraterrestre A.C." (IMMEU A.C. in Spanish) a civil organization which looks to contribute to the development and innovation of aerospace technology and its derivatives (spin off). Among its main projects are the development of new generation liquid fuel sounding rockets.[71] The INMEU together with a new company called Datiotec Aeroespacial and the Autonomous University of San Luis Potosí (UASP in Spanish) obtained funding from the AEM-CONACYT funding program in 2015 for a project to develop an auxiliary propulsion system for rockets.[72] Datiotec Aeroespacial also developed training platform consisting of picosatellites called CanSat.[73] and is part of a group

---

69 ENMICE, "ENMICE website", ENMICE 2023. https://enmice.mx/.

70 Mario Manuel Arreola Santander, "Lanzadores Espaciales Hechos en México", Hacia el Espacio magazine, 2 February 2020. https://haciaelespacio.aem.gob.mx/revistadigital/articul.php?interior=1082.

71 Museo del Espacio de San Luis potosí, "Cohetes de Nueva Generación", Museo del Espacio de San Luis Potosí hospedado en la Facultad de Ciencias de la Universidad Autónoma de San Luis Potosí. http://galia.fc.uaslp.mx/museo/espacio/9.htm.

72 Fondo Sectorial de Investigación, Desarrollo Tecnológico e Innovación en Actividades Espaciales, "Desarrollo de un sistema de propulsión auxiliar para cohetes", Fondo sectorial AEM-CONACYT. https://conahcyt.mx/wpcontent/uploads/convocatorias/fondos_sectoriales_constituidos/aem_conacyt/fichas_de_conclusion/FICHAS_AEM_CONVOCATORIA_2014/AEM_249078.pdf.

73 LATAM Satelital, "Datiotec Aeroespacial de México desarrolla satélites de bajo costo", Tecnología Espacial, LATAM Satelital, 11 April 2016. https://www.cienciamx.com/index.php/tecnologia/tic/5628-cuantumlabs-una-empresa-de-desarrollo-tecnologico-de-vanguardia.

of three companies.[74] It is unclear the current status of both, INMEU and Datiotec Aeroespacial.

Although several of the mentioned events require the investment from public funds through AEM itself or CONACYHT, few of these could obtained public funding and in most of the cases, the organizations have operated based on their own funding or private donations apart from government institutions-based events, whose funding comes from their annual budget allocation.

Special mention should be made of the International Astronautical Congress (IAC) held in Mexico for the first time in 2016. The 67th IAC edition was held in Guadalajara, Mexico, under the motto: “Making space accessible and affordable to all countries”. The AEM believed that bringing the IAC in 2016 to this region of the world would contribute positively and give an additional boost to the efforts of involved emerging nations into a new era of international collaboration to solve common problems using space science and technology.[75] With a participation of 5217 people; 60 entities were present on the exhibition floor in 15,000 square meters, where meetings, exhibitions, encounters, and more were held. The 67th IAC 2016 had the largest media coverage, with 200 professionals attending 150 simultaneous meetings. On the last day, the end of the Rosetta Probe Mission was broadcasted, which took place at 5:30 a.m. Mexico time, marking the end of twelve years since its liftoff and more than 20 years of preparation.[76]

5.2 Small Satellites and Robotic Missions from Mexican Institutions

By the end of the 2010s, several Mexican universities succeed in launching CubeSats (a nanosatellite with a size and cube form factor of 10 cm created in 1999)[77] with all satellite missions based on technology demonstration payloads and capacity building. Table 4 shows the nanosatellites design, developed, launched and operated by Mexican entities at the time of this publication.[78] The funding for all satellites came from both, public and private institutions as well as the Mexican government thorough the AEM-CONACYT funding program for space science

[74] Roxana de León, “CuantumLabs: innovación en electrónica”, Cienciamx noticias. https://www.cienciamx.com/index.php/tecnologia/tic/5628-cuantumlabs-una-empresa-de-desarrollo-tecnologico-de-vanguardia.

[75] International Astronautical Federation, Events. https://www.iafastro.org/events/iac/iac-2016/.

[76] Evangelina del Toro, “México cumple con sus primeros Juegos Olímpicos del Espacio”, Agencia Espacial Mexicana & NOTIMEX, 2 October 2 2016. https://aem.gob.mx/iac2016-NOTIMEX/notas/247913.html.

[77] Ravi A. Deepak, Robert J. Twiggs, *Thinking Out of the Box: Space Science Beyond the CubeSat*, Journal of Small Satellites, 2012. https://www.jossonline.com/wp-content/uploads/2014/12/0101-Thinking-Outside-the-Box-Space-Science-Beyond-the-CubeSat.pdf.

[78] Valeria Bereniz Ramos Barba, *An overview of the current state of national legislation, policy, and regulation in Mexico*, UNISEC-Global The 43rd Virtual Meeting. https://unisec-global.org/pdf/virtual-meeting/2024/43rd/43rd_Virtual_UNIGLO_report.pdf.

Table 4 Nanosatellites developed by Mexican entities and their launch by chronological order

Satellite name	Developers	Funding	Facts	Launching date
Painani-1 	SEDENA, CICESE	SEDENA, AEM	3U CubeSat First Mexican nanosatellite launched into orbit Launched by Electron rocket (RocketLab)	26 June 2019
AztechSat-1 	UPAEP	AEM, NASA-AMES	1U CubeSat First Mexican nanosatellite launched from ISS	5 December 2019
Nano Connect-2 	LINX-UNAM	CITNOVA Hidalgo, small and medium enterprises (SMEs)	2U CubeSat First nanosatellite completely developed in Mexico (as indicated by the developers) Launched from India	27 February 2021
Painani-2 	SEDENA, CDA-IPN, PEU-UNAM, FI-UNAM	SEDENA, AEM	3U CubeSat Launched from Cape Canaveral by Falcon 9 rocket	30 June 2021

and technology (which was dismissed in 2020 by Presidential Order)[79] as well as collaboration work between AEM and NASA in the particular case of the AztechSat-1 nanosatellite.[80]

There are other nanosatellite projects that are either, under development or funding which are summarized in Table 5.[81] Several of the listed projects in Table 5 have faced funding problems creating a situation in which some of them were suspended or in the worst case, have not yet formally started.

Regarding robotic mission, it is notable the project COLMENA, which was the first Mexican mission to the Moon launched on 8 January 2024. The missions are comprised of a set of five self-organizing microrobots that will collect data from the lunar surface, a complicated environment exposed to the interplanetary

[79] Diario Oficial de la Federación (DOF), "DECRETO por el que se ordena la extinción o terminación de los fideicomisos públicos, mandatos públicos y análogos", 2 April 2020. https://www.dof.gob.mx/nota_detalle.php?codigo=5591085&fecha=02/04/2020#gsc.tab=0.

[80] Carlos Duarte, *AztechSat1 a model for cooperation in space development*, United Nations-Brazil Symposium on Basic Space Technology, Creating Novel Opportunities with Small Satellite Space Missions, 2018, p. 5. https://www.unoosa.org/documents/pdf/psa/activities/2018/Symposium_Brazil_BSTI/presentations/oral/S13_01_UNBSTI208_CarlosDuarte.pdf.

[81] Valeria Bereniz Ramos Barba, "An overview of the current state of national legislation, policy, and regulation in Mexico", UNISEC-Global The 43rd Virtual Meeting. https://unisec-global.org/pdf/virtual-meeting/2024/43rd/43rd_Virtual_UNIGLO_report.pdf.

Table 5 Nanosatellites under development or funding by Mexican entities and their launch by chronological order

Satellite name	Developers	Funding	Facts	Launching date
K'OTO 	UAT-FI UNAM	Querétaro government, UNAM	1U CubeSat Technology demonstration and human resources development	TBD
PAKAL (Colibrí mission) 	Panamerican University (UP)	Panamerican University (UP)	3U CubeSat Debris assessment in LEO and human resources development	TBD
GXIBA-1 	UPAEP	UPAEP	1U CubeSat Monitoring of active volcanoes in Mexico Launch through Kibo CUBE (JAXA)	TBD
IXAYA 	PEU-UNAM	TBD	3U CubeSat Forest and Mexico City environment monitoring	TBD
AztechSat Constellation 	UNAM, UNAQ, UP, UPQ, UPAEP	TBD	The project consists of designing, building, testing and operating at least four CubeSats for animal telemetry applications, i.e., to track and monitor marine species	TBD

medium. With a diameter of 12 cm and a weight of less than 60 g, the dimensions of these robots will place them just a few centimetres from the lunar regolith surface. Regolith is a Moon indigenous powder made up of extremely fine, irregular and abrasive grains.[82] Although the COLMENA mission could not reach the Moon due to a propellant leak with the Lunar Lander (Peregrine 1 spacecraft),[83] the mission completed at least 50% of success when it was turned on and operated for a limited time set by the Lunar Lander due to power budget constraints under

[82] Laboratorio de Instrumentación Espacial LINX, "COLMENA", Instituto de Ciencia Nucleares UNAM. https://linx.nucleares.unam.mx/colmena-en/.

[83] NASA Space Science Data Coordination Archive, "Peregrine Mission 1 (Astrobotic)", NASA. https://nssdc.gsfc.nasa.gov/nmc/spacecraft/display.action?id=PEREGRN-1.

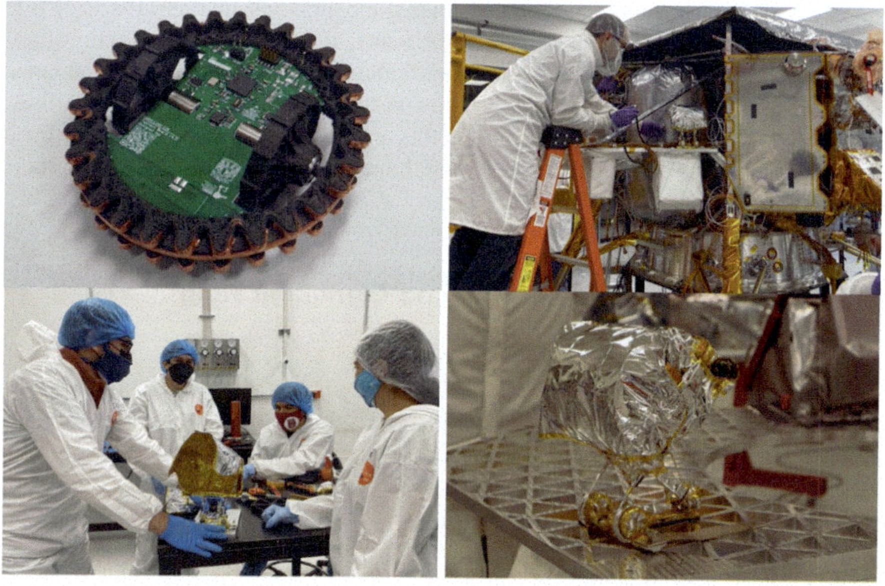

Fig. 4 COLMENA mission robots and integration with Lunar Lander

the failure situation. The 75% of success was indicated to be completed some hours later after the first successful operation.[84] The Fig. 4 shows the COLMENA mission.[85]

6 The Challenges to Consolidate the Space Activities in Mexico in the Long Term

Space activities in Mexico face several unique challenges, including technological, economic and regulatory. National space capabilities can span the entire space value chain ranging from (i) manufacturing of various types (satellites, launch vehicles, components), (ii) ground infrastructure and (iii) downstream services and applications. A good strategy will be to determine through which activities can generate competitive advantages, depending on the resources available and the

[84] Boletín UNAM-DGCS-020, "Misión COLMENA: avances cruciales en el espacio profundo", UNAM Global, 2024. https://unamglobal.unam.mx/global_revista/mision-colmena-avances-cruciales-en-el-espacio-profundo/.

[85] Laboratorio de Instrumentación Espacial LINX, "Proyectos COLMENA", LINX Instituto de Ciencias Nucleares UNAM. https://linx.nucleares.unam.mx/proyectos/.

objectives that each country sets in its plans and policies in this area.[86] While Mexico is part of international space agreements, further integration and collaboration with global space agencies and private companies are needed to advance its capabilities. In this section we will focus on the challenges regarding satellites, rockets and other activities to consolidate the space sector in Mexico. As we have seen throughout this historical journey, there have been some steps and the impetus to promote these areas.

6.1 Satellite Regulations Beyond Communication Applications

As indicated before, in the PSGF there is an extensive and explicit mention of satellites applied to communication applications, giving them a clear priority for the Mexican government. For instance, the only mention to small satellites appears in Section 2 (The Background) where it is indicated that "*Mexico has academic institutions and satellite applications, in particular in the field of small satellites.*"[87] This indicates that the PSGF focus on satellite communications, understanding as satellites those operating in GEO in a "traditional" way: satellite with mass of > 1000 kg and operating for several years. This is more evident in point 5 of the PSGF where it is mentioned that "*In recent years, traditional GEO satellite providers have increasingly partnered with new providers in Low Earth Orbit (LEO) and Medium Earth Orbit (MEO), as well as with traditional terrestrial telecommunications providers, to combine technologies to serve customers in different coverage areas. However, it is estimated that the traditional global telecommunications infrastructure will continue to have much more capacity (2000 terabits per second (Tbit/s) than the projected space infrastructure capacity (estimated at 50 Tbit/s by 2026).*"[88]

On the other hand, the Regulation of the Federal Telecommunications and Broadcasting Law on Satellite Capacity as a state reserve, published on 25 April 2024, considers mainly GEO satellites by explicitly indicating the GEO orbital positions and the frequency bands assigned by the International Telecommunications Union (ITU) along the text.[89] It refers only to other orbits different from GEO in terms of LEO to an orbit from 400 km to 2000 km, and medium orbits from

[86] BryceTech, "Global Space Strategies and Best Practices", p. 3. https://www.space.gov.au/sites/default/files/mediadocuments/202311/global_space_strategies_and_best_practices_-_research_paper_0.pdf.

[87] Diario Oficial de la Federación (DOF), "ACUERDO que establece la Política en materia satelital del Gobierno Federal", 2023. https://dof.gob.mx/nota_detalle.php?codigo=5712152&fecha=21/12/2023#gsc.tab=0.

[88] Ibid.

[89] Diario Oficial de la Federación (DOF), "DECRETO por el que se expide el Reglamento de la Ley Federal de Telecomunicaciones y Radiodifusión en Materia de Capacidad Satelital como Reserva del Estado", DOF, 25 April 2024. https://dof.gob.mx/nota_detalle.php?codigo=5724805&fecha=25/04/2024#gsc.tab=0.

the LEO to the GEO, which left the orbit section where a considerable portion of satellites operate at some point in their lifecycle and even from the beginning of operation when deployed from the International Space Station (in the case of small satellites): the LEO orbit below 400 km. This situation becomes more complex when considering satellites and spacecrafts operating beyond GEO such as the Lunar Lander that took the COLMENA mission in its fly to the Moon orbit, which is a fact currently ongoing and is set to increase near in the future.

Even though satellite industry (applications and their associated ground equipment) represents around three quarters of the total global space economy,[90] the satellite applications consider, besides communications (television, telephone, broadband, emergency communication, maritime, road and rail) the following:

- Remote Sensing

  a. Agriculture, forest and fisheries monitoring
  b. Environmental protection
  c. Oceanography
  d. Cartography & landscape change
  e. Disaster mitigation
  f. Civil infrastructure
  g. Meteorology
  h. National resources exploration and utilization

- Scientific Research and Space Science

  a. Space weather
  b. Astronomy and astrophysics
  c. Earth sciences studies (atmosphere, geology, ice coverage, oceans, etc.)
  d. Education and amateur satellites

- Navigation

  a. Location tracking
  b. Timing and synchronization
  c. Military operations

- National Security

  a. Maritime domain awareness
  b. Surveillance and reconnaissance
  c. Space Situational Awareness (SSA).

[90] Bryce, “2022 Global Space Economy at a Glance”, Bryce Reports, 2022. https://brycetech.com/reports.

The void of laws and regulations regarding the other satellite applications decreases Mexico's competitiveness in providing such services (satellites that are operated within the country as well as those that offer services to users domestically). Furthermore, the lack of legal certainty in this area limits the attraction of foreign investment as well as the creation of domestic companies.

Although several satellites have been launched in Mexico to date, in addition to the projects currently under development, their regulation is still deficient. There is a guide prepared by the Centre for Scientific Research and Higher Education of Ensenada, Baja California (CICESE) called, Regulatory Orientation Guide for Small Satellites, which is a reference for developers in Mexico and considers the main aspects such as[91]:

- Procedure to obtain the concession of orbital resources (frequencies)
- Registration of space objects
- Import–export processes
- Space debris avoidance considerations.

While the referred Guide provides basic information on how to operate a small satellite developed in Mexico, there are regulatory challenges that must be addressed by the Mexican government. Among these challenges are[92]:

- Licensing: it should consider the capacity and activity of the intended satellite to avoid undesired activities, or by uncertainty inhibit the development of others. Routinary launches of small satellites will require ease of licensing process to avoid bottle neck situations.
- Registration: many satellites are not registered, which puts their identification at risk in case of any accident or conflict.
- Frequency resources regulation: small satellites use mostly amateur radio bands (BRA), which currently causes saturation of these bands.
- Space debris: the space debris guidelines are not legally binding (mandatory) under international law and must be implemented through national mechanisms. The domestic application of these is a challenge in Mexico as there are no national regulatory mechanisms that limit the issue of licenses for such satellites.
- Liability: the liability for damage caused by the operation of small satellites will impose regulatory and financial challenges, requiring a balance in the form of

[91] CICESE, "Guía de orientación regulatoria para satélites pequeños no sujetos a regulación", Fondo Sectorial de Investigación, Desarrollo Tecnológico e Innovación en Actividades Espaciales CONACYT—AEM, 2015. https://www.educacionespacial.aem.gob.mx/images/normateca/pdf/GOR_11_LR.pdf.

[92] Valeria Bereniz Ramos Barba, Isai Fajardo Tapia, "Satélites pequeños", NOTA-INCyTU, Number 038, March 2020. https://foroconsultivo.org.mx/INCyTU/documentos/Completa/INCYTU_20-038.pdf.

appropriate technical and regulatory measures without letting uncertainty limit their development.

6.2 Transition from Experimental/Amateur to Space Capable Rockets

In Mexico experimental rocketry is developing again after the first stage during 1959 to 1975 under SCT and CNEE as mentioned before. Amateurs with some certifications from international associations were participating with well-elaborated rocket models in national and international competitions; however, despite having their designs validated in commercial software, quality manufacturing and successful launches, significant gaps impede progress towards development of professional rockets.[93]

> Amateur rocketry in Mexico faces several challenges, such as the lack of economic, material and human resources; the scarcity of adequate spaces for launches; the difficulties to obtain the necessary permits; and the lack of visibility and social recognition. This activity is a way for space development in Mexico, as it contributes to generate knowledge, talent and innovation in the field of propulsion.[94]

The AEM is the primary government body responsible for coordinating Mexico's space activities, including the promotion of space research and development. However currently lacks a detailed regulatory framework specifically for space vehicles like rockets.

> Experimental rocketry aims at education, learning and advancing technological development, while professional rocketry must adhere to local and national regulations for rocket launches. This includes detailed mission planning, covering aspects such as logistics, safety and environmental compliance. In addition, professional rocketry involves the mass production of rocket engines and vehicles, which implies large-scale manufacturing with the aim of generating commercial operations.[95]

As indicated by one of its most active amateur rocketry members in recent years, Luis Arellano:

---

[93] Pablo Alejandro Arizpe Carreón, Revista Electrónica Hacia el Espacio, AEM, Transición de los Cohetes Experimentales a los Profesionales: Aspectos Clave para México, 2 July 2024. https://haciaelespacio.aem.gob.mx/revistadigital/articul.php?interior=1584.

[94] Carlos Duarte, Revista Electrónica Hacia el Espacio, Agencia Espacial Mexicana, La cohetería amateur en México: Un impulso para el desarrollo espacial, 27 November 2023. https://haciaelespacio.aem.gob.mx/revistadigital/articul.php?interior=1533.

[95] Pablo Alejandro Arizpe Carreón, Revista Electrónica Hacia el Espacio, AEM, Transición de los Cohetes Experimentales a los Profesionales: Aspectos Clave para México, 2 July 2024. https://haciaelespacio.aem.gob.mx/revistadigital/articul.php?interior=1584.

> From the beginning we have encountered several difficulties to develop our activities, such as the lack of regulation of space activities. In Mexico, there is a long-standing absence due to the lack of activity in this field, which is why there is a lack of public policies to promote the development of the sector.[96]

Currently, there is no specific Mexican law that comprehensively regulates the development, launch, and operation of space vehicles like rockets. The legal framework is primarily focused on general aerospace activities, scientific research, and the use of explosives, leaving significant gaps in the regulation of space vehicle technology.

Regarding to adequate spaces Luis Arellano says that "…safe locations are required for flight tests and static tests, which do not pose a risk to people and facilities. Mexico has sufficient territory for these activities, but there is not yet a defined area for them."[97] Mexico lacks the extensive space infrastructure that more established spacefaring nations have. This includes launch facilities, research institutions, and manufacturing capabilities.

> The space industry requires special materials and alloys, as well as instruments, advanced equipment and components that are difficult to obtain, and which are often not catalogued in the national customs system for import, which complicates and hinders their entry into the country.[98]

From geopolitical perspective, Mexico's space aspirations are influenced by its relationships with the United States of America, one of the main space-capable nations, which can affect trade of advanced components, technology transfer and collaborations.

Pablo Arizpe indicates that:

> It is crucial to strengthen certain aspects to ensure a successful transition, these include the need to improve project management, systems engineering, the development of procedures, training programs and budget control. In addition, detailed planning for laboratories, test benches and specialized equipment is essential. These efforts will ensure an effective transition to professional rocketry in Mexico.[99]

From the Mexican Public Policy perspective, the PNAE 2018–2024[100] establishes as part of Priority Objective Number 3.1. "*Promote the development of own*

[96] Mario Arreola, *Lanzadores Espaciales Hechos en México,* Revista Electrónica Hacia el Espacio, AEM, 2 April 2020. https://haciaelespacio.aem.gob.mx/revistadigital/articul.php?interior=1082.

[97] Ibid.

[98] Ibid.

[99] Pablo Alejandro Arizpe Carreón. Revista Electrónica Hacia el Espacio, AEM, Transición de los Cohetes Experimentales a los Profesionales: Aspectos Clave para México, 2 July 2024. https://haciaelespacio.aem.gob.mx/revistadigital/articul.php?interior=1584.

[100] Mexican Government, AEM, "Programa Nacional de Actividades Espaciales 2020–2024", AEM. https://www.gob.mx/aem/documentos/programa-nacional-de-actividades-espaciales-2020-2024.

capacities to achieve technological independence in space exploration for the benefit of Mexicans". Regarding to rocketry, the unique specific action is the item 3.1.4 "*Establish collaboration programs through alliances with higher education institutions, research centres and the private sector to promote rocket design and development activities*". In this sense, there is an effort by the government entity AEM in alliance with local governments for the creation of the two Regional Centre for Space Development (CREDES), one is in Atlacomulco, in the State of Mexico and the other one in the City of Zacatecas.

In the case of the CREDES-Atlacomulco, the construction began in January 2019 by the Ministry of National Defence in a facility consisting of two main buildings. The built-up area covers 5000 m^2, consists of a satellite design and development laboratory, a rocket design and development laboratory and a digital image processing laboratory. It also has a Space Technology Transfer Office and human capital training areas.

The CREDES-Atlacomulco has the following specific objectives[101]:

- Carry out innovation, technological development and applied research projects.
- Transfer and democratize space knowledge and technologies.
- Offer access to infrastructure and services.
- Offer educational and training services.
- Promote the development of industry and space innovation projects.
- Contribute to the creation of technical regulations for the space sector in Mexico.

For its part, CREDES-Zacatecas,[102] seeks to transform research and technological development in telecommunications and electronics for the space sector. Among the strategic objectives of CREDES-Zacatecas, the following stand out:

- Increase R&D projects in space telecommunications.
- Spark the development of specialized human capital in the areas of electronic telecommunications, mechatronics, energy and satellite technologies.
- Development of an ecosystem of start-ups, spin-offs and innovation processes.

At CREDES-Atlacomulco, activities aimed at the design and development of rockets are carried out, including analysis and modelling of components or elements of rocket engines, as well as launch vehicles. The laboratory has the following specialized software: ANSYS Mechanical and ANSYS Fluent; Solid Works, Container for engine testing.

[101] Mexican Government, AEM, CREDES-Atlacomulco. https://aem.gob.mx/credes/EdoMex.html.

[102] Mexican Government, AEM, CREDES-Zacatecas. https://aem.gob.mx/credes/Zac.html.

CREDES research centres are seen as multidisciplinary organizations, and therefore aim to generate products for a broad spectrum of potential users, including public institutions of the federal, local and municipal; national and international companies; higher education institutions; and in general, any institution with objectives that benefit the Mexican population, especially those that need it most.[103] Both centres started operations on 2022, at present they have published as mentioned above the objectives; also, the infrastructure, services and projects, that need a sufficient investment to develop own technical capabilities. Mexico still lacks the extensive space infrastructure that more established spacefaring nations have, despite the efforts mentioned above.

6.3 The New Structure of the Space Related Organisms in Mexico: The Creation of the Digital Transformation and Telecommunications Agency

On 28 November 2024, the new president of Mexico, Dr. Claudia Sheinbaum Pardo, created by decree and published on the DOF, the Digital Transformation and Telecommunications Agency (Agencia de Transformación Digital y Telecomunicaciones or ATDT in Spanish language),[104] a new entity at the level of the other secretaries of state and whose main work focuses on the following 7 main objectives[105]:

1. To make existing face-to-face procedures available digitally.
2. Minimising the regulatory burden on individuals and businesses.
3. Generate savings and opportunity for zero corruption.
4. Create the Centro de Atención para el Bienestar (CABI), through the 079 number, which provides citizen attention 24 h a day.
5. Implement intelligence and data analysis to strengthen public capacities.
6. Technological autonomy and information security.
7. Internet as a right.

The new agency is proposed to be built over two main organisational pillars to achieve the mentioned objectives:

- **Digital Transformation**: which will be dedicated to:

[103] Ibid.

[104] Diario Oficial de la Federación, "DECRETO por el que se reforman, adicionan y derogan diversas disposiciones de la Ley Orgánica de la Administración Pública Federal", 2024. https://www.dof.gob.mx/nota_detalle.php?codigo=5744005&fecha=28/11/2024#gsc.tab=0.

[105] Mexican government press release, 14 November 2024, https://www.gob.mx/presidencia/prensa/presidenta-claudia-sheinbaum-presenta-la-nueva-agencia-de-transformacion-digital-para-simplificar-tramites-y-evitar-la-corrupcion.

(a) the simplification and digitalisation of procedures with the MX Key,
(b) to the management of the new Welfare Attention Centre, which is the single attention number;
(c) to the technological assessment which will be in charge of the approval of contracts to avoid abuses, cost overruns and to ensure that resources can be reused.

- **Digital Public Infrastructure**: the enabling area in charge of executing:

(d) software manufacturing
(e) technology infrastructure
(f) data centres and analytics
(g) cybersecurity
(h) telecommunications policy.

Under the new ATDT agency, the Mexican government decided to simplify several procedures and cover a wide range of different areas. This drastic change includes moving the following four parastatal entities and one deconcentrated administrative body under the administration of the new agency[106]:

- Servicio Postal Mexicano (Mexican Postal Service),
- Organismo Promotor de Inversiones en Telecomunicaciones (Telecommunications Investment Promotion Agency),
- Agencia Espacial Mexicana (AEM),
- INFOTEC Centro de Investigación e Innovación en Tecnologías de la Información y Comunicación (Centre for Research and Innovation in Information and Communication Technologies),
- Comisión Nacional de Mejora Regulatoria (National Commission for Regulatory Improvement).

Besides the movement of the mentioned entities, all the telecommunications attributions under the decentralised body called 'Financiera para el Bienestar', will also be transferred to the ATDT, which includes all the human, material and financial resources consisting of the Telecommunications network and Mexsat; and the transfer of the human, material and financial resources currently available to the Directorate of the Trunk Network of the Federal Electricity Commission.

It is important to mention that among all the different attributions that the new ATDT entity will have, those related to the space activities at the time of the present publication, and explicitly mentioned in the decree of November 2024, are:

[106] Diario Oficial de la Federación, "DECRETO por el que se reforman, adicionan y derogan diversas disposiciones de la Ley Orgánica de la Administración Pública Federal", 2024. https://www.dof.gob.mx/nota_detalle.php?codigo=5744005&fecha=28/11/2024#gsc.tab=0.

1. To elaborate and conduct the Federal Government's telecommunications and broadcasting policies.
2. To define the satellite capacity that, where appropriate, is required of persons holding concessions for orbital resources and those authorised to exploit the rights to broadcast and receive signals from foreign satellites to provide services in the national territory, as a State reserve for national security networks, social services and other government needs, and monitor the efficient use of their own satellite capacity, whether granted or acquired, or that established as a State reserve.

It is clear that the telecommunications regulation and all its related procedures will be managed by the ATDT, however it is uncertain how the rest of the space activities and their regulation will be managed. This initiative preserves the Mexican government's biased vision of the space sector, which, far from strengthening it, it keeps it fragmented and diffuse.

6.4 Other Activities Necessary for the Consolidation of the Space Sector in Mexico

It is essential that the constitutional reform proposal be approved to consider space activities at the constitutional level and generate the National Law on Space Activities and secondary regulation of the sector. The above for the purpose to provide legal certainty and comply with the obligations arising from international treaties on outer space of which Mexico is a party.

FAMEX is an important event that takes place every two years and has shown significant growth in its almost ten years of activity. It has the potential to bring companies from the global space sector to Mexico's space ecosystem, which is why it is important to provide an adequate legal framework for the development of national space activities that considers the worldwide space situation.

The investment in space R&D is low, which hinders innovation and the development of cutting-edge technology. The highest budget allocated to the AEM was M.X.N.$ 112,000,000 equivalent to U.S.$ 6,200,000 in 2014. After all, the average budget to date is around U.S.$ 3,000,000 per year. On the other hand, the spending on R&D in Mexico is 0.27% of GDP for 2022,[107] below the average of 2.95% by the OECD members during 2020.[108] In this sense, it will be necessary to increase public spending on R&D and promote private investment in the space sector.

CONACES has positioned itself as an important meeting forum for engineers, researchers, practitioners, companies, and representatives from the industry and the Government. However, it requires continuous support from the government and

[107] World Bank, *Research and development expenditure (% of GDP) - Mexico*. https://datos.bancomundial.org/indicador/GB.XPD.RSDV.GD.ZS?locations=MX.
[108] World Bank, *Research and development expenditure (% of GDP) - OECD members*. https://datos.bancomundial.org/indicador/GB.XPD.RSDV.GD.ZS?locations=OE.

industry to maintain its operation and consolidate its relevance and impact. Only five people are part of the CONACES committee as indicated on its website, so the continuous allocation of more human and financial resources, as well as sponsorships are needed to ensure the continuity of this important national congress. Expanding its scope to become international would help to position the research space activities in Mexico in the international community.

The President-Elect of Mexico for the period 2024–2030, proposes in her project "100 steps for the transformation of Claudia Sheinbaum"[109] as step 79 the launching of a Mexican satellite into orbit that will contribute to communications. However, this proposal is not clear. At present Mexico has no capacity to develop and launch its satellites making it dependent from other countries which means it would be desirable for the technology transfer to occur in the next generation of GEO satellites that might be the scope of the mentioned step 79. In this sense, the PNAE 2018–2024[110] establishes as its 2040 Vision that the country will have the capacity to manufacture telecommunications and Earth observation satellites, with a minimum of 50% of national components. It will also develop launch services for light payloads to medium and LEO orbits, which will allow the country to put the corresponding space services into operation.

In relation to the objective of increasing the country's capabilities and promoting cooperation in science and technology, in space exploration for the scientific and technological strengthening of Mexico, special emphasis is placed on Action Proposal 3.1.2 on "*Taking advantage of the intellectual capital of Mexicans abroad who are developing space science and technology to support local developments.*" In this regard, it is proposed to create participation and consultation mechanisms through which Mexicans abroad can express their concerns and proposals to improve public policies in space matters and their regulation. In terms of cooperation, to publicize and make easily accessible the calls for proposal, projects and channels for collaboration. For example, a mechanism could be created in which the Institute for Mexicans Abroad (IME in Spanish) participates in coordination with AEM-CONAHCYT-IME for that purpose; in the case of CONAHCYT through RedCyTE or similar.

7 Conclusions

It is essential that the national constitutional reform to be approved so that Mexican outer space activities are considered priority areas. This will enable the creation of the National Law on Outer Space Activities and Secondary Regulation of the sector.

It is important to foster a better regulatory environment for outer space sector in Mexico to promote competitiveness, quality, innovation and technological

[109] Claudia Sheinbaum, *100 pasos para la Transformación*, Movimiento de Regeneración Nacional (MORENA) 2024. https://claudiasheinbaumpardo.mx/wp-content/uploads/2024/03/CSP100.pdf.
[110] PNAE 2028–2024, p. 54.

development, which will provide new opportunities to enter international trade and business.

There is a national lack of laws and regulations on all space activities, with efforts focused only on satellite communications applications. There is also a lack of a long-term vision of the future of satellite applications in general and of their funding: keeping the isolation and privilege of satellite communications over other satellite applications compromises the PSGF itself. Furthermore, small satellites are not relevant in the current satellite policy, although they represent a more accessible approach to growing technical capabilities in the field of satellites and even though several Mexican small satellites have been launched in recent years.

The AEM is the primary government body responsible for coordinating Mexico's space activities, including the promotion of space R&D. However, currently it lacks a detailed regulatory framework specifically for space vehicles like rockets. The Mexican legal framework is primarily focused on general aerospace activities, scientific research, and the use of explosive materials, leaving significant gaps in the regulation of space vehicle technology.

At present Mexico has no capacity to develop and launch own satellites, making it dependent from other countries which means it would be desirable for the technology transfer to occur in the next generation of GEO satellites.

At present, the operational status of the RedCyTE is not clear. The efforts to create this network with a national and international presence should not be discarded. It is recommended that it will be strengthened since it represents an academic arm that pushes and covers all areas of outer space knowledge, thus having impact on the design and implementation of public policies based on science and technology.

Figure 5 shows a synthesis of the main activities of public and private institutions, and the reached milestones from the Government perspective (orange text), as well as the domestic laws and regulations on which the space activities are currently based on. The blue text is related to the Constitutional Amendment is pending on approval at the time of this publication.

In summary, Mexico space development achievements started with a big enthusiasm simultaneously as other space powers started too and continued mainly with the construction and operation of ground infrastructure for satellite communications for several decades. Space research and exploration gained a second boost with the creation of the AEM in 2010: in the following few years there have been a variety of domestic space activities ranging from hosting the International Astronautical Congress (IAC) in 2016, launching several small satellite missions from universities and governmental institutions, launching a mission to the Moon, the creation of local amateur rocket competitions, the creation of domestic technical standards, and the proliferation of more university programs in the aerospace field. The academic, the public, partially the Government and in smaller scale the industry, have shown interest in the space activities in and for the country. However, more effort is required from the Government to provide necessary laws and regulation for the players, and from the industry to promote business and domestic investment from both, national and international entities. So, the Mexican space ecosystem will become a more relevant future orientated actor that contributes to

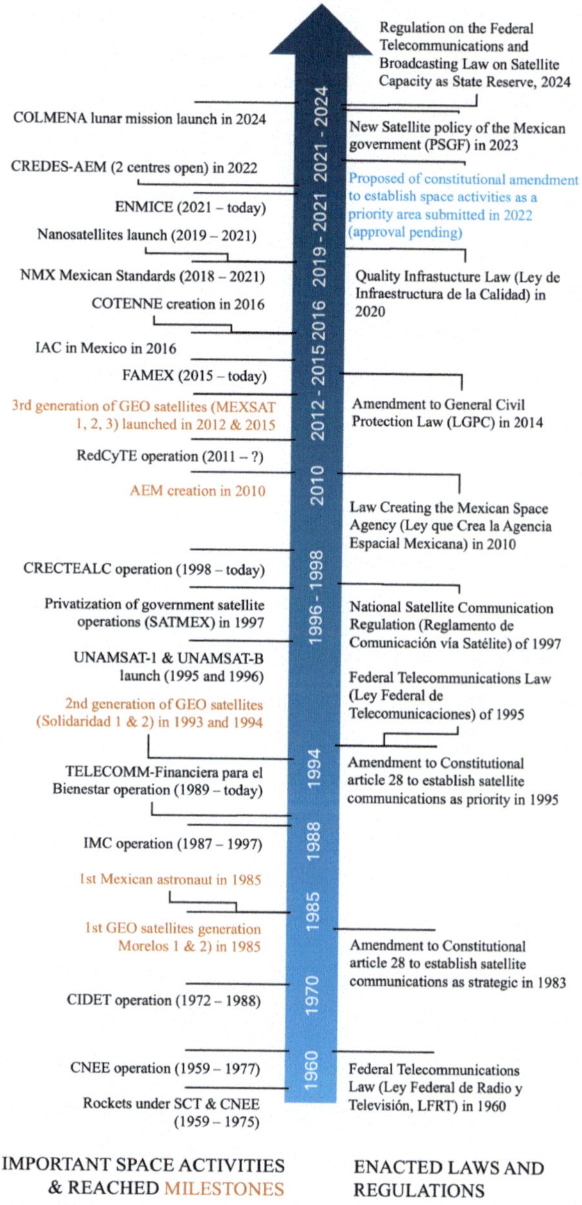

Fig. 5 Important space activities and reached milestones, enacted laws and regulations of Mexico since 1960

solve the problems of the country, as well as the human space activities and their challenges in the years to come.

Valeria Bereniz Ramos-Barba is a Space Law and Policy Consultant. She received her master's in law from the University of Granada, Spain. She is a member of the Global MX Network, Japan Chapter of the Institute for Mexicans Abroad. She has participated and organized academic events to make proposals on public policies, legislation and regulation of space activities in Mexico. Her current research areas include international development, space sustainability and national framework for space activities in emerging countries.

The Potential of Mexico's Capabilities in Utilizing Severe Plastic Deformation to form Al-Li Parts for Aerospace Applications

Bárbara Bermúdez-Reyes, J. A. Ferrer-Pérez, and Rafael Vargas-Bernal

Abstract

Weight optimization of aircraft structures requires materials that offer the required performance using lightweight components primarily based on sustainable and recyclable aluminum alloys. This material offers low densities compared to other materials, thus reducing the aircraft's weight and increasing load capacity and fuel efficiency. Hardness and strength are improved by combining alloying and heat treatment to withstand high loads and vibrations. In aerospace applications, where harsh environments are present, properties such as fatigue strength and corrosion resistance must be improved to extend the service life of components. Severe plastic deformation of aluminum alloys allows the production of materials with unique property-microstructure relationships capable of offering high-strength and other functional properties by exploiting the phase transformations of materials under mechanical loading. The primary purpose of this chapter is to examine Mexican infrastructure and the technological capabilities to develop components based on aluminum-lithium alloys through severe plastic deformation for aerospace applications. However, despite the advances, there is a pressing need for increased scientific research that promotes the technical capabilities of manufacturing aircraft components based on these materials in Mexico.

B. Bermúdez-Reyes (✉) · J. A. Ferrer-Pérez
Advanced Technology Unit, School of Engineering, UNAM, Juriquilla Querétaro, Mexico
e-mail: barbara.bermudez@ingenieria.unam.edu

J. A. Ferrer-Pérez
e-mail: ferrerp@unam.mx

R. Vargas-Bernal
Departamento de Ingeniería Automotriz, Universidad Politécnica de Guanajuato, Guanajuato, México
e-mail: rvbernal@upgto.edu.mx

J. A. Ferrer Pérez and A. Froehlich (eds.), *Space Fostering Mexican Society*, Southern Space Studies, https://doi.org/10.1007/978-3-031-92617-4_2

1 Introduction

The rich history of aerospace development in Mexico, which began shortly after the Second World War, is a testament to the country's potential in this field. This historical journey, marked by technological development activities and constant innovation since 1957, not only from an academic and governmental point of view but also from a business point of view in areas such as manufacturing, electronics, metalworking, and device assembly, is a source of pride and connection for all involved. This has generated industrial ecosystems associated with opening new companies (nearshoring), increasing technological development and product design, and introducing and updating study plans.[1]

The aerospace industrial sector is based on companies dedicated to manufacturing, maintenance, repair, and operations services, engineering and design, and auxiliary services for civil and military aircraft. It is considered that these industries in Mexico will establish themselves as one of the aerospace industry corridors in North America. In addition to experiencing constant growth, this is evidenced by the fact that in 2022, 300 million dollars were generated, which represented 0.6% of Mexico's domestic product.[2] From this perspective, it is expected that 2030 the aerospace industrial sector will reach 500 million dollars, which will significantly favor the national economy.

An essential sector of the aerospace industry is manufacturing metal parts (metal-mechanical area), which provides casting, forming, machining, and production services for parts made of various alloys of aluminum, titanium, nickel, and steel, among others. This implies that Mexico must have a complete knowledge of national and international legislation, instructions, procedures, and standards.[3]

Therefore, specialized personnel are necessary in metallurgy, materials, and aerospace manufacturing. According to recent data, the Mexican aerospace industry requires around 40,000 engineers trained in metalworking.[4] This was made clear during the Aerospace Meeting 2024, which emphasized the opportunity and responsibility for Mexico to remain at the forefront of technology and production, implying that aerospace engineers must be trained in manufacturing, materials, and assembly areas.

[1] Secretaría de Economía, "Plan de Órbita 2.0. Mapa de Ruta del Sector Espacial Mexicano", (Ciudad de México: PROMEXICO, 2017).

[2] Aeroespacial, "Industria aeroespacial mexicana: en cifras", *Metalmecánica,* 2023, https://www.metalmecanica.com/es/noticias/industria-aeroespacial-mexicana-en-cifras (all websites cited in this publication were last accessed and verified on 5 Agoust 2024).

[3] Dianny Niño, "La apuesta de México por el 10 de la arena global aeroespacial", 2023, https://www.metalmecanica.com/es/noticias/la-apuesta-de-mexico-por-el-top-10-de-la-arena-global-aeroespacial.

[4] Paola Castellanos, "Industria aeroespacial requiere suplir demanda de 40 mil ingenieros", *Metalmecánica,* 2024, https://www.metalmecanica.com/es/noticias/industria-aeroespacial-requiere-suplir-demanda-de-40-mil-ingenieros.

Furthermore, at the same event, it was reported that Mexico will need approximately 40,000 engineers in the aerospace area by 2035.[5] Table 1 presents the aerospace engineering programs that are currently available in México. Note that a close relationship exists between academia and the private sector, which has intensified with the aerospace industry growth. This is due to the rapprochement of both branches when the academic programs are proposed or updated. This closes the gap between the real needs of the aerospace sector and the contribution to national development, even though Mexico only invests 0.5% of its Gross Domestic Product in research and development.

Table 1 Aerospace engineering programs in Mexico

Institution	Program	Specialty	State
Universidad Autonoma de Baja California[6]	Aerospace Engineering	Design	Baja California
Universidad Autonoma de Chihuahua[7]	Aerospace Engineering	Aerospace Design Manufacturing and production	Chihuahua
Universidad Popular Autónoma del Estado de Puebla[8]	Aerospace Engineering	Structural Design Manufacturing Process	Puebla
Universidad Marista de Guadalajara[9]	Aerospace Engineering	Structural design Propulsion	Jalisco
Universidad Nacional Autonoma de México[10]	Aerospace Engineering	Aeronautics Astronautics	Mexico City and Queretaro
Universidad Internacional de Innovación[11]	Aerospace Engineering and Space Exploration	Aeronautics Astronautics	Aguascalientes
Universidad Aeronautica en Querétaro[12]	Aerospace Engineering Master	Propulsion Manufacturing Aerospace security	Queretaro

[5] MMS México, "En Yucatán destacan capacidades de la industria aeroespacial", *Modern Machine Shop Mexico*, 2021, https://www.mms-mexico.com/noticias/post/en-yucatan-destacan-capacidades-de-la-industria-aeroespacial/.

[6] UABC, "Ingeniero Aeroespacial", *Universidad Autónoma de Baja California*, 2020-I, http://web.uabc.mx/formacionbasica/FichasPE/Ingeniero_Aeroespacial.pdf.

[7] UACH, "Ingeniero Aeroespacial", *Universidad Autónoma de Chihuahua*, 2024, https://uach.mx/pregrado/ingeniero-aeroespacial/.

[8] UPAEP, "Ingenieria Aeroespacial", *Universidad Popular del Estado de Puebla*, 2024, https://upaep.mx/licenciaturas/ingenieria-aeroespacial.

[9] UMG, "Ingenieria Aeroespacial", *Universidad Marista de Guadalajara*, 2024, https://umg.edu.mx/portal/ingenieria-aeroespacial/.

[10] UNAM, "Ingenieria Aeroespacial", *Universidad Nacional Autónoma de México*, 2023, https://www.ingenieria.unam.mx/programas_academicos/licenciatura/aeroespacial.php.

[11] UNII, "Ingenieria Aeroespacial y Exploración del Espacio", *Universidad Internacional de Innovación*, 2024, https://unii.edu.mx/wp-content/uploads/2024/07/ING-AEROESPACIAL-Y-EXPLORACION-DEL-ESPACIO.pdf.

[12] UNAQ, "Maestría en Ingenieria Aeroespacial", *Universidad Aeronáutica en Queretaro*, 2024, https://www.unaq.edu.mx/oferta-educativa/posgrado/mia/.

Therefore, the academy is also a critical factor in the supply chain if the demand for aerospace engineers in nearshoring with the United States, Canada, Korea, China, and Europe has become more evident.[13] To the above, PROMEXICO and the Ministry of Economy published in 2017 the "Roadmap of the Space Sector in Mexico," which includes the aerospace area. This document proposes four thematic axes that make up the sector's strategy as follows:[14]

- *Axis 1*: Innovation and niche opportunities for the industry and convergent services. It consists of the development of regulatory frameworks, exploration, and industrial and business facilities,
- *Axis 2*: Self-determination of the development of space activities and cooperation for strengthening the Mexican sector. Access to aerospace technology, orbital positions, development, flight, and control of aircraft,
- *Axis 3*: Boosting the sector's value chain and promoting industrial development. It focuses on the development of industrial capacities and infrastructure to actively participate in the international aerospace market,
- *Axis 4*: Promoting digital access and development of applications and solutions. It seeks to increase connectivity throughout the country, search and exploitation of content, and online backup applications.

For all the above, this chapter describes Mexico's capabilities, advances, and advantages in manufacturing applied to the aerospace sector and the feasibility of manufacturing parts for high-speed aircraft.

It is for these reasons that one of the intentions of the Mexican government and, according to the Ministry of Economy for 2035, Mexico must serve the various markets that make up the aerospace sector, in addition to positioning the country among the leaders of the Global Share Market.

2 Advances in Aerospace Manufacturing in Mexico

Mexico is an important manufacturing hub for aerospace parts. It has 350 international companies in the aerospace sector. It is the 14th largest aerospace supplier at the world level, and it contributes approximately 80% of production exported to the United States and the rest to Canada, France, and Germany.[15] After Brazil,

[13] Adriana Cao García, "Nota Sectorial: La industria Aeroespacial en Mexico", *Antena Igape Mexico*, 2021, https://igape.gal/images/05-mais-igape/05-05-quensomos-internacional/antenas/NotaSectoria_LaIndustriaAeroEspacialenMexico_032021.pdf.

[14] PROMEXICO, *Plan de Órbita 2.0, Mapa de Ruta del Sector Aeroespacial Mexicano*. (México, Secretaria de Economía -PROMEXICO, 2017).

[15] Diana Armero Somolinos, "Aeroespacial en México", *ICEX-España Exportación e Inversiones*, 2023, https://www.icex.es/content/dam/es/icex/oficinas/077/documentos/2023/12/fichas-sector/FS_Aeroespacial%20en%20M%C3%A9xico%202023_REV.pdf.

Table 2 Value chain and aerospace products

Value chain	Products
Engineering and design	Balloons, airships, wings, gliders, unpowered and powered aircraft Helicopters, airplanes, spacecraft, launch vehicles, and suborbital vehicles
Manufacturing and maintenance	Parts of the engineering and design value chain Parts and accessories for electrical and electronic machines and devices
Repair and reconditioning	Maintenance, repair, and overhaul (MRO) of fixed-wing and rotary-wing aircraft

Mexico is the second largest producer in Latin America in terms of manufacturing parts for aircraft, due to: [16]

1. There is an increase in passenger demand, and it is expected that by 2032, it will double,
2. Defense spending: the budget for military use and the purchase of military aeronautical products has increased.

Therefore, one of the Mexican government's most relevant strategies is to facilitate the entry of foreign companies and support the national education system in training highly specialized human resources in aerospace manufacturing and materials.

According to the National Institute of Statistics and Geography (INEGI), Mexico has:

- 1st place in goods for assembly and manufacturing of aircraft parts,
- 2nd place in exports of parts for turbojets or turboprops.
- 3rd place in manufacturing of parts for gas turbines.

This is why the Ministry of Economy divides the Mexican aerospace industry into three value chains (Table 2):

Based on Table 2, Mexico's aerospace industry is primarily located in TIER1 and TIER2. This shows that Mexico is a high-quality manufacturing country that complies with international and national aerospace regulations. For this reason, the Ministry of Economy classifies the aerospace industry in global value chains, as presented in Table 3.

Therefore, Mexico represents many industrial opportunities because it is located in North America. Having qualified labor specialized in the aerospace sector allows

[16] Mordor Intelligence, "Análisis de tamaño y participación en México- Tendencias de crecimiento y pronósticos (2023-2024)", *Mordor Intelligence*, 2024, https://www.mordorintelligence.com/es/industry-reports/mexico-data-center-market.

Table 3 Mexican aerospace value global chain

Global value chain	Value chain	Products
Original equipment manufacturer	Ingeneering and design Manufacturing and maintenance	Final products: engines, airplanes, helicopters, and satellites
TIER1 (systems)	Manufacturing and maintenance Repair and reconditioning	Final assembly line: interiors, wings, landing gear, fuselage, turbines, etc.
TIER2 (subsystems)	Manufacturing and maintenance	TIER1 Suppliers: Harnesses, Subassemblies, Rings, etc.
TIER3 (sub tier)	Manufacturing and maintenance	Connectors, cables, insulating materials, composites, laminates and switches

Mexico to generate income and improve the population's quality of life by offering better services and increasing the country's quality of life.

3 Advantages for the Aerospace Sector in Mexico

The market analysis carried out by Mordor Intelligence regarding growth trends and forecasts for 2024–2029 indicates that in Mexico, the aerospace industry will be 22.70 million dollars by 2029, which implies an annual rate of 15.18%.

During the COVID-2019 pandemic, the Mexican aerospace industry was affected because exports decreased by more than 20%, which affected national income. Currently, the aerospace sector has been strengthened due to the low production costs of the metalworking sector, which is very high in production.

Therefore, the aerospace sector ranks Mexico as the fourteenth aerospace supplier, and Mordor Intelligence's prediction for 2024-2029 indicates market volume growth in megawatts (MW), as shown in Fig. 1. This market volume growth measure refers to the electrical consumption that the industry uses for manufacturing components, which also indicates the growth and updating of the aerospace infrastructure in Mexico.

Concerning the above information, Mexico represents an excellent opportunity for:[17]

- Import substitution in the North American region,
- Independence from China in terms of operations,
- Conversion and transformation of small to medium-sized companies and their integration into the high-value global chain sectors of the aerospace sector.

[17] Mauricio Pineda, "Industria Aeroespacial en Mexico: Oportunidades y nuevas fronteras", *Modern Machine Shop Mexico,* 2021, https://www.mms-mexico.com/articulos/industria-aeroespacial-en-mexico-oportunidades-y-nuevas-fronteras.

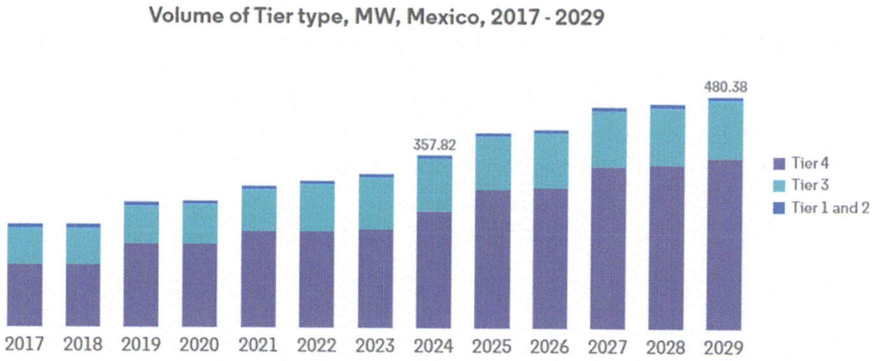

Fig. 1 Market volume growth Aerospace industry (2017–2029)

The main advantage of the aerospace industry in Mexico is the metalworking and materials industry, which creates high-performance alloys for aerospace industry parts. According to Raúl Pérez-Bustamante of Modern Machine Shop México, there is an interest in strengthening this sector so that, in 2025, global investment reaches 430.87 billion dollars, which will be reflected in more significant opportunities, technological challenges, and new methods of manufacturing materials.[18] This leads to new research being proposed in Mexico related to technological development and innovation in the aerospace sector, which will allow for an increase in the qualified and highly specialized workforce at all training levels (technicians, engineers, and postgraduates) in the disciplines of materials, metallurgy, and manufacturing.

Another advantage is that specialized clusters in the aerospace sector bring together large, medium, and small companies and academic institutions with programs focused on Aerospace, materials, and metallurgy.[19] Currently, in Mexico, six aerospace clusters are presented in Table 4.

Since there are few aerospace clusters, studies are being carried out to locate companies in the aerospace sector that are not registered in any cluster. More than 350 companies across 19 states, as shown in Figure 2. Table 5 shows the number of companies and their locations.[20]

[18] Raúl Pérez Bustamante, "Aleaciones de alto desempeño para piezas de la industria aeroespacial", *Modern Machine Shop Mexico*, 2022, https://www.mms-mexico.com/articulos/aleaciones-de-alto-desempeno-para-piezas-de-la-industria-aeroespacial.

[19] Christian Muñoz-Sanchez, María del Rocio Soto-Flores, Luis Rocha-Lona, and Jose Arturo Garza-Reyes, "Aerospace Industry in Queretaro, Mexico: A perspective of Regional Innovation System", *Proceedings of the International Conference on Industrial Engineering and Operations Management Bangkok*, (Thailand, March 5-7, 2019).

[20] Miguel Angel Vazquez and Carmen Bocanegra, "The Aerospace in Mexico: Characteristics and Challenges in Sonora", *Revista Latinoamericana de Economía* 49, no. 195 (2018).

Table 4 Mexico´s Aerospace clusters

Cluster	Location
Baja Aerospace Cluster	Baja California
Cluster de Energía Sonora	Sonora
Chihuahua Aerospace Cluster	Chihuahua
Monterrey Aerocluster	Nuevo León
Aerocluster Queretaro	Queretaro
Cluster Aeroespacial del Bajio	Guanajuato

Fig. 2 Location of the Mexican aerospace industry

Table 5 Aerospace companies and their location

State	Companies	State	Companies	State	Companies
Baja California	97	Guanajuato	14	Durango	3
Sonora	58	Ciudad de Mexico	12	San Luis Potosí	2
Chihuahua	52	Tamaulipas	12	Puebla	2
Querétaro	50	Yucatán	12	Zacatecas	2
Nuevo León	33	Jalisco	10	Aguascalientes	1
Coahuila	14	Estado de México	10	Hidalgo	1
				Oaxaca	1

Fig. 3 The virtuous circle of how the aerospace industry operates in Mexico

According to Jesús Castillo, this location of the aerospace industry is generating business networks and agglomerations with the country's sectoral production network. This has a direct impact on the hiring of personnel who are being trained and specialized, which is an excellent advantage because it eliminates recertification costs by foreign governments and simplifies customs costs and procedures.[21]

So, in numbers: Mexico has more than 350 companies in the aerospace sector, of which 80% are involved in aircraft parts manufacturing, 11% in maintenance and repair, and 9% in engineering and design. These companies represent 1/8 of the country's workforce. Figure 3 shows a diagram that shows a virtuous circle of how the aerospace industry operates in Mexico.

All of the above is also reflected in the increase in workers' skills and salaries, which also generates an increase in competitiveness due to the quality of the workforce. As the Mexican aerospace industry evolves and grows, as well as the relationships that have been generated with academia, government, and society, the Government of Mexico includes Materials Science, Aerospace, and Aeronautical Engineering on the list of national priority areas for solving problems through

[21] Jesús Castillo Rodríguez, "Globalization and localization of the aerospace industry in Mexico", *Intersticios Sociales* 26, (2024), 181.

science, technology, and innovation. In addition, the Mexican Airspace Protection Law will be enacted in 2023.[22]

Due to the above advantages, aerospace activities in Mexico have generated and grown many opportunities for growth at all levels, in addition to facilitating the incubation of start-ups that provide specific and specialized aerospace services, such as precision agriculture, aerial photography, cartographic updating, monitoring of water basins, forests, and jungles, design and manufacturing of multirotor and drones, etc.[23] This allows the country to be competitive in areas such as manufacturing and generating materials, providing air services, and continuing to be a strategic aerospace hub in North America.[24]

4 Lithium deposits in Mexico

The history of lithium in Mexico dates back to the late 1980s when research and exploration of its discovery began in Sonora, Puebla, and Oaxaca. 2008, the private sector restarted this project, but its exploration began in Baja California, San Luis Potosí, and Zacatecas (Fig. 4).[25]

As a result of the previous explorations, 9049100 tons of lithium were found available in hard rock, brine, and clay deposits. On the other hand, according to the studies carried out by the General Directorate of Mining Development and private exploration companies in 2020, 43.3 million tons of lithium with a grade of 0.30% lithium were estimated, which placed Mexico in 10th place among countries with lithium deposits. According to these data, there are eight states in the country with lithium reserves, in which 57 deposit locations were found, and 11 were considered viable (Fig. 5).[26]

The above data will allow Mexico to research, develop technology, and innovate in the energy, medicine, and aluminum metallurgy sectors for electrical, automotive, and aerospace applications.[27]

[22] Consejo Nacional de Ciencia y Tecnología, "Programa Institucional 2020–2024 del Consejo Nacional de Ciencia y Tecnología", *Diario Oficial de la Federación*, 2020, https://dof.gob.mx/nota_detalle.php?codigo=5595309&fecha=23/06/2020#gsc.tab=0.

[23] Jose Carlos Rodríguez, María Aline Manzo and Mario Gomez, "High-Performance Level and Innovation Capabilities in the Aerospace Industry of Mexico: A Qualitative Comparative Analysis", *International Journal of Innovation and Technology Management* 20, no. 3 (2023): 2350011.

[24] María Aline Manzo and José Carlos Rodríguez, "Ensuring the High Performance of Design and Engineering Firms in Mexico's Aerospace Industry: A Qualitative Comparative Analysis", *International Journal of Technology* 13, no.1 (2022): 5–12.

[25] Ricardo Agustín Sánchez Mancera and Roberto Pérez Garibay, "El litio de México", *Revista Avance y Perspectiva CINVESTAV* 9, no. 3 (2023).

[26] Aleida Azamar, "El litio en México: verdades y mentiras", *Minería en México: panorama social, ambiental y económico,* (2022): 27-45.

[27] Bryan W. Jaskula, "Lithium", *US Geological Survey-Mineral Commodity Summaries*, 2024, https://pubs.usgs.gov/periodicals/mcs2024/mcs2024-lithium.pdf.

Fig. 4 Exploration of lithium deposits in Mexico

Fig. 5 Lithium deposits were found in Mexico until 2020

Table 6 Activities to be developed for lithium exploitation

Activity	Forms of development
Engineering, research, geological activities, exploitation, benefit, and use of lithium	Direct development
Production, transformation, and distribution of lithium-derived products	Partnerships with public and private institutions

Currently, no lithium deposit is being exploited, and progress has been made in forming a public company for its exploitation. This is because lithium was nationalized in 2022. However, lithium is part of the constitutional strategic materials; therefore, Mexico's exclusivity or reserve will be maintained.[28]

On the other hand, the lithium value chain includes exploitation of deposits until the marketing of the final product. Therefore, this activity has become economically strategic and greatly impacted Mexico and other international markets.

Therefore, Mexico by becoming an economic agent and following the decree of nationalization of lithium, the Ministry of Economy was appointed as the one in charge of managing and controlling the productive chains of the economic value of lithium. Therefore, the classification of activities that must be developed (Table 6).

The above leads to updating and regulating both the mining and industrial sectors. Therefore, it will be essential to consider private initiatives for extracting lithium and its use in the various production sectors. This would give Mexico a competitive capacity due to its geographical location (access to the Gulf of Mexico and the Pacific Ocean) and the Mexico-United States-Canada Treaty (T-MEC), in which Mexico could also provide lithium and enter the value chain.[29]

If all of the above is realized, lithium would allow the metalworking and aerospace sectors to develop even more due to the:

- Metalworking sector: Lithium decreases the melting point and viscosity of the aluminum metal bath and increases electrical conductivity and elasticity,
- Aerospace sector: Aluminum-lithium alloys present resistance to tension in fastening elements (bolts and rivets) and increase the load capacity of an aircraft.

[28] José Roldan Xopa, "Litio para México: Estudio sobre la empresa pública y la regulación del litio en México", *Natural Resource Governance Institute*, (2023).

[29] Oscar Ocampo, "¿México como potencia del litio?", *Instituto Mexicano de la Competitividad y Centro de Investigación en Política Pública*, 2022, https://www.imco.org.mz/mexico-como-potencia-de-litio.

5 Innovation in Aerospace Manufacturing in Mexico

The application of lightweight materials and the design optimization allow aircraft parts to be produced, which means that new technologies are used in conventional manufacturing. This encourages staff to update and specialize even more.[30]

Furthermore, this has contributed to the significant reduction of fuels and the carbon footprint. The signing of the T-MEC has also contributed to the government's flexibility in public policies, reflected in government support to attract foreign investment, the generation of jobs, and the contribution to the country's gross domestic product.

The National Chamber of the Aluminum Industry was also created in 2017, which capitalizes on the Mexican market and has achieved in-country production, creating 122,500 direct jobs and more than 1 million indirect jobs.[31]

Therefore, Mexico has become one of the primary recipients of investment in aerospace manufacturing worldwide and is the country that has been involved in research and technological development projects for various aircraft components. This has led to technological developments in Mexico, such as manufacturing aluminum structural parts for high-speed aircraft.

Currently, in Mexico, research and characterization of Al-Li alloys has begun. Since it is not a conventional alloy, it requires non-conventional metallurgical processes. Because lithium is a highly reactive element due to its low molecular weight, to be added, a high level of control over the Al impurities and an Argon atmosphere must be used to protect it from possible explosions.[32] The fusion process must be carried out by electrolysis, casting in metal molds, then undergoing a homogenization treatment and continuing with the secondary manufacturing processes.[33]

However, up to this point, the microstructure has been anisotropic, which makes the forming processes complex and generates part waste due to fractures and tears caused by the heterogeneous distribution of the alloying elements and the formation of compounds in the microstructure.[34] Table 7 shows the alloying elements and impurities that comprise the Al-Li alloys and their effects.[35]

---

[30] Aeroespacial, "Avances en la industria aeroespacial mexicana 2024", *Metalmecánica*, 2024, https://www.metalmecanica.com/es/noticias/avances-en-la-industria-aeroespacial-mexicana-2024.

[31] Secretaría de Economía, "Resolución definitiva por la que se autoriza la constitución de la Cámara Nacional de la Industria del Aluminio", *Diario Oficial de la Federación*, 2016, https://www.dof.gob.mx/nota_detalle.php?codigo=5436734&fecha=11/05/2016#gsc.tab=0.

[32] Ian Polmear, "Light Alloys. From Traditional Alloys to Nanocrystals -Fourth Edition", (Oxford: Butterworth Heinemann, 2006), 436.

[33] Adrian P. Mouritz, "Introduction to Aerospace Materials", (Cambridge: Woodhead Publishing, 2012), 636.

[34] Roberto J. Rioja, and John Liu, "The Evolution of Al-Li Base Products for Aerospace and Space Application", *Metallurgical and Materials Transactions A* 43A, (2012): 3325–3337.

[35] Olga Grushko, Boris Ovsyannikov, and Viktor Ovchinnikov, "Aluminum Lithium Alloys: Process Metallurgy, Physical Metallurgy", (Boca Raton: CRC Press, 2017), 1.

Table 7 Alloying elements and impurities that make up Al-Li alloys and their effects

Alloying elements and impurities	Effects
Li and Mg	Density reduction, formation of solid solution, and consolidation of precipitate formation
Cu	Formation of solid solution and consolidation of precipitate formation
Ag	Nucleating agent, precipitate former of the Al_2CuLi phase, increasing the cost of the alloy
Zn	Consolidation of the solid solution and provides corrosion resistance
Zr, Sc, Mn, Cr	Formation of dispersoid compounds and control grain size and microstructural texture
Ti and B	Grain refiners
Fe and Si	Impurities that form intermetallic compounds and reduce metallic properties
Na and K	Impurities

The impurities described above form a group of compounds, and their effects are described in Table 8:

For the above, in Al-Li alloys, the chemical composition and proportions of the alloying elements and the process temperature must be considered. Since they are alloys with an anisotropic microstructure, it is difficult to predict the failures that

Table 8 Description of compound groups in Al-Li alloys

Compounds groups	Compounds types	Effect
Al-Li	δ'(Al_3Li) and AlLi-Mg y Al_2LiMg precipitates	They restrict shear stresses, hinder dislocation movements, and prevent grain coarsening during precipitation
Al-Cu	θ'(Al_2Cu) and Mg precipitates (Guinier-Preston Zone)	They promote nucleation of the Guinier-Preston zone in the octahedral voids of the crystal lattice
AlCuLi	Al_6CuLi_3, Al_2CuLi y la fase Al_7Cu_4Li o $Al_{7.5}Cu_4Li$ Precipitates	They affect toughness and are formed by peritectic reaction and excess Mn
Intermetálicos	AlCuFeSi, Al_7Cu_2Fe, Al_3Fe, AlCuFeMn	They affect precipitation kinetics
AlCuMg	Al_2Cu Mg y σ($Al_5Cu_6Mg_2$)	They affect precipitation kinetics
AlCuMgAg	Ω(Al(cu-Ag) (Li-Mg)) y γ'($AlAg_2$) precipitates	They accelerate precipitation kinetics
Dispersoids	Al_3Sc, Al_6Mn, $Al_{20}Cu_2Mn$, β'(Al_3Zr) and core-shell compounds (Al_3Sc/Al_3Zr)	Grain coarsening harms recrystallization and texture control during solidification or aging at high temperatures

occur during manufacturing (tears, transverse fractures, and grooves).[36] It should be noted that the same anisotropy does not allow recrystallization heat treatments to refine the grains because Al-Li alloys degrade thermally.[37]

Therefore, the SPD process has been used in Mexico to refine the microstructure at room temperature since 2010.[38] The severe plastic deformation-forming technique of materials is based on the Hall-Petch relationship:[39]

$$\sigma_y = \sigma_0 + Ad^{\frac{1}{2}} \tag{1}$$

where σ_y is the yield strength, σ^0 is the initial stress, A is the material constant, and d is the mean grain diameter. Equation (1) states that the smaller the grain size, the more excellent the resistance to deformation. Therefore, the SPD process is very efficient in refining the grain size and thus obtaining an isotropic material and obtaining specific properties and three conditions must be met:[40]

- High working pressures,
- The dimensions of the piece must not present significant changes,
- Significant reduction in grain size.

To achieve the above, the SPD process has six techniques, of which there are:

- Angled channel extrusion (ECAP),
- Sheet metal banding (ARB),
- High-pressure torsion (HPT),
- Iterative corrugation and straightening (RCS),
- Cyclic closed die forging (CCDF),
- Super short multi-pass rolling (SSMR).

With the SPD process, the grain size can be refined to less than 1 μm, which increases the tensile strength without reducing toughness, allowing for obtaining

[36] R. Sepe, V. Giannella, N. Razavi, and F. Berto, V., "Characterization of static, fatigue, and fracture behavior of the aluminum-lithium alloy Al Li 2198 T851, *International Journal of Fatigue* 166, (2023): 107265.

[37] A.A. Mazilkin, and M.M. Myshlyaev, "Microstructure and thermal stability of superplastic aluminum-lithium alloy after severe plastic deformation", *Journal of Materials Science* 41, (2006): 3767.

[38] Edgar Alberto Ortiz Cuellar, "Efecto y comparación de técnicas de deformación plástica severa sobre la microestructura y propiedades mecánicas en una aleación de aluminio 6060", Master Thesis, (Monterrey: UANL, 2010).

[39] A. Azushima, R. Kopp, A. Korhonen, D.Y. Yang, F. Micari, G.D. Lahoti, P. Groche, J. Yanagimoto, N. Tsuji, A. Rosochowski, and A. Yanagida, "Severe plastic deformation (SPD) processes for metals", *CIRP Annals Manufacturing Technology* 57, no. 2 (2008): 716–735.

[40] Jorge Mauricio Cubero Sesin, "Desarrollo de Aleaciones nanoestructuradas de Aluminio con Propiedades Mecánicas Superiores Mediante Deformación Plástica Severa", *Revista Universitaria* 16, no.9 (2015) : 1–13.

parts that will present high dispersion and transmission of stresses through the structure. The ECAP technique is the most widely used, in which a single channel is subjected to angular extrusion (90° extrusion). Being an angled extrusion, a shear deformation is generated. If performed several times or several passes, a material with ultra-fine grains with isotropic properties, high mechanical and corrosion resistance, and no changes in the geometry and dimensions of the material can be achieved.[41]

Therefore, these processing and conditioning characteristics of Al-Li alloys using the ECAP technique could be indicated for obtaining wing and fuselage spars for high-speed aircraft made of Al-Li alloys, and according to Polmear and Mazilkin, the following characteristics can be obtained:

- 10 to 12 % lighter than conventional Al alloys,
- Grain size reduction to 0.7 μm,
- 3 % reduction in density,
- The elastic modulus increases by 5 %,
- Increase in tensile strength,
- Increase in corrosion resistance.

For this reason, the ECAP technique improves the mechanical properties of Al-Li alloys at room temperature.

High-speed flights are carried out in aircraft whose structure is very stable and has a high structural safety factor. The speeds reached by these aircraft are more significant than the speed of sound:

- Supersonic flights: $1.2 < M < 5$,
- Hypersonic flights: $5 < M$,

where M is the Mach number (Eq. 2), which is a relative speed generated between the speed of a body (V) and the speed of sound in the medium in which the body moves (a).

$$M = \frac{V}{a} \tag{2}$$

Due to the speeds handled, the structures with the most significant mechanical demands are the wing and the fuselage. For the latter, the alloy must resist fatigue and cracking because the spars and frames of the structure are subjected to tension, compression, radial, and longitudinal stresses. On the other hand, being exposed to atmospheric contaminants (sulfides, chlorides, and dioxins), the material must

[41] Jinghua Jiang, Ting Yuan, Wuwen Zhang, Aibin Ma, Dan Song, and Yuna Wu, "Effect of equal channel angular pressing and post aging on impact toughness of Al-Li Alloys", *Materials Science and Engineering: A* 527, (2018): 4716.

resist dynamic and stress-assisted corrosion.[42] For the wing structure, being subjected to straight and level flight, acrobatic and turns, the structure is subjected to combined stresses (tension, compression, bending, rotation, and torsion) and vibration; consequently, the material must present high resistance to mechanical stresses, vibrations, and high fracture toughness.

For all the above reasons, Mexico can form Al-Li parts through severe plastic deformation for high-speed aircraft structures. In addition, the metalworking sector is considered a long-term sector, as it shows continuous growth and supports the aerospace sector.[43]

6 Conclusions

Mexico has many advantages. Its evolving aerospace and metalworking industrial sectors mean it has the capacity and confidence to manufacture and shape Al-Li alloys. It also has educational programs at all levels in aerospace engineering, metalworking, materials, and manufacturing. Although more human resources will be required in the future, the current nearshoring will ensure that all commercial, industrial, and professional demands are met.

Acknowledgments The authors are grateful for their valuable financial support to the Secretariat of Science, Humanities, Technology, and Innovation (SECIHTI) Project CBF2023-2024-333.

Bárbara Bermúdez-Reyes Received a Ph.D. in Metallurgy and Materials Science in 2009. She is a Professor at the High Technology Unit of the National University Autonomous of Mexico. Previously, she was a Research Professor at different universities. She was General Secretary of the University Space Engineering Consortium-Mexico Chapter for 2014–2019. She is a member of the Space Science and Technology Network. She has collaborated on projects with the Mexican Air Force and has evaluated projects of the Mexican Naval Research Institute. She is part of the National Research System with level I. Her research lines are designing and processing aerospace ceramic-ceramic and metal-ceramic composite materials, satellite structures, and severe plastic deformation to form aluminum alloy for high-velocity aircraft.

Jorge Alfredo Ferrer-Pérez is a National University Autonomous of Mexico-School of Engineering professor. He received his Ph.D. in Aerospace and Mechanical Engineering from the University of Notre Dame, South Bend, in the United States. He is part of the Aerospace Engineering Department and is responsible for the Space Propulsion and Thermo-vacuum lab. This facility belongs to the Space and Automotive Engineering National Laboratory located at Juriquilla. His

[42] Chunhua Ran, Miao Ding, Xiaofeng Li, Yiqing Ge, Xiaofang Xiao, Le Shen, Fan Wang, Jingfeng Wang, and Hong Chang, "Experimental study on the mechanical behavior of 5A90 aluminum-lithium alloy subjected to combined tensile shear loadings The combined strength of 5A90 aluminum-lithium alloy", *Materials Science and Engineering* 527, (2023): 4716.

[43] Mauricio Pineda, "Aeroespacial, una industria de largo plazo", *Modern Machine Shop México*, 2017, https://www.mms-mexico.com/articulos/aeroespacial-una-industria-de-largo-plazo.

current research areas are nano-heat transfer in solid-state devices, thermal control, space propulsion, small satellites, and space technology development.

Rafael Vargas-Bernal received a bachelor's degree in communications and Electronics Engineering from the University of Guanajuato in 1995, Mexico, and the degrees of Master of Science and Doctorate in Science with a Specialty in Electronics from the National Institute of Astrophysics, Optics and Electronics (INAOE) in 1997 and 2000 from Mexico, respectively. Since September 2024, he has been a professor at the Polytechnic University of Guanajuato, where he established himself as a researcher in electronics and materials engineering. He has authored 19 articles in journals, 50 chapters in books, and more than 100 conference articles. He has edited three books for IntechOpen. He is a member of the National System of Researchers (SNI-Mexico). He regularly serves as a reviewer of scientific articles in RSC, Elsevier, MDPI, Wiley, Taylor & Francis, and ACS, as well as industrial standards in Semiconductor Equipment and Materials International (SEMI). His research interests include two-dimensional materials, nanomaterials, space materials, composite materials, gas sensors, and biosensors.

Uses and Applications of Artificial Intelligence and Its Contribution to the Growth of the Mexican Space Industry

Castillo Malagón Romina, Farah-Simón Lisette, and Buenrostro Aguilar Hugo Javier

Abstract

The space industry has the potential to generate growth and economic development, as well as contribute to the solution of problems worldwide; in recent years it has had unprecedented growth, its exploration ranges from academic and research purposes to the search for new business models such as obtaining resources or space tourism. Among its particular characteristics are the development and adoption of high-tech goods, including those associated with artificial intelligence (AI); which can encourage the development of industry in developing countries. For the Mexican context, the adoption of AI in the space industry can favor the fulfillment of the objectives proposed in the National Program of Space Activities 2020–2024 of the Mexican Space Agency as well as benefit other sectors such as construction, tourism and agribusiness. Therefore, the objectives of this research are introduced in Part 1, which are to identify the main applications of AI in the space sector, as well as its contribution to the fulfillment of the National Program of Space Activities 2020–2024 of the Mexican Space Agency. To this end, Part 2 briefly presents the contributions of the theory of economic change, whose main contribution is the use of technology to improve production processes and generate growth and economic development. Part 3 defines, what AI is and, describes some of its components. General space sector applications are addressed, thus leading to reflect viable aspects which impact Mexico and its National Program of Space Activities 2020–2024. In Part 4, a literature review is carried out and the information obtained through the creation of knowledge networks are analyzed, then, secondary sources of information were consulted to identify technological trends associated with

C. M. Romina (✉)
Instituto Politécnico Nacional, Mexico City, México
e-mail: rocastillom@ipn.mx

F.-S. Lisette · B. A. H. Javier
Universidad Nacional Autónoma de México, Mexico City, México

J. A. Ferrer Pérez and A. Froehlich (eds.), *Space Fostering Mexican Society*, Southern Space Studies, https://doi.org/10.1007/978-3-031-92617-4_3

AI that impact the space sector and that can also be used in the Mexican case, especially in sectors such as tourism, agro-industrial, manufacturing and construction. Finally, some conclusions are presented.

1 Introduction

Despite the recognition by the international community of the existence of airspace and outer space or space, no consensus has been reached on the physical delimitation between them. In this sense, with the intention of regulating its operation, applicable regulations and activities to be developed, some technical or scientific criteria have been established to discriminate airspace from space. Such criteria are the composition of the atmosphere, the density of the air as an element of support for aircraft, the delimitation considering the orbit of an artificial satellite of the Earth, the Von Karman Line (in 100–118 km altitude) and the delimitation based on the effects of gravity.[1] In this sense, the most accepted division refers to the Karman Line, marking the beginning of space between 100 and 118 km above sea level.

There are different motivations for the exploration of space, ranging from the search for knowledge to the exploration of new resources and spaces to inhabit outside the planetary orbit, as well as the use of new technologies and innovative solutions to address global problems on planet Earth.[2] As space evolves towards the search for new business models, the possibility of exploiting natural resources in situ, ranging from water to rare elements such as Helium-3, which are relevant for energy production, becomes relevant.[3] The possibility of using space for tourism purposes is also being considered, with the provision of building space hotels and making (sub)orbital trips.

According to the report issued by the World Economic Forum and the international consulting firm McKinsey and Company, the space sector is experiencing unprecedented growth, where the development and adaptation of technology are playing a key role. This can be seen in the testing of new rocket systems to pioneering satellite launches and complex robotic missions to the Moon and beyond. In this sense, it is expected that by 2035 the activities associated with space exploration and exploitation will be worth USD 1.80 trillion compared to the USD 63 trillion generated during 2023.[4]

[1] Calvo Castillo, M. Ángel, (2020), Airspace and its international regulations. Iuris Tantum, 34(31), pp. 111–124. https://doi.org/10.36105/iut.2020n31.09.

[2] Johnson-Freese, J., (2007), Space as a Strategic Asset. Columbia University Press. https://doi.org/10.7312/john13654.

[3] Dolman, E., (2002), Astropolitik Classical Geopolitics in the Space Age. Routledge.

[4] World Economic Forum and McKinsey and Company, (2024), Space: The $1.8 Trillion Opportunity for Global Economic Growth.

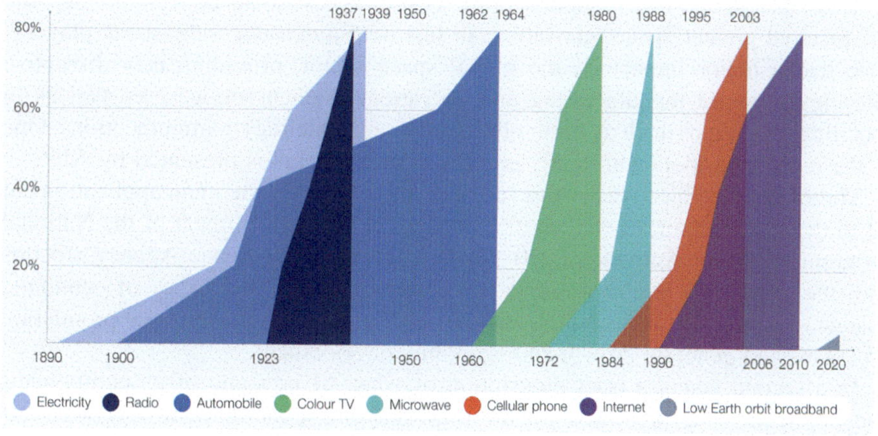

Fig. 1 Dissemination of innovative products (penetration rate in U.S. households, %). *Source* US Census Bureau; The New York Times; Enjeux-Les Echos quoted in WEF and McKinsey and Company (2024)

This set of activities and the use of resources that create value and benefits for human beings in the course of exploring, researching, understanding, managing and utilizing of space is defined by the Organization for Economic Cooperation and Development[5] as space economy.

The space economy is growing and evolving, along with the development and profound transformation of the space sector and the greater integration of space into society and economy. Examples of this are the technology and infrastructure developed to strengthen space activities, and which is used in other sectors such as meteorology, energy, telecommunications, insurance, transport, maritime transport, aviation and urban development, which translates into additional economic and social benefits. Therefore, it is considered as a sector that promises great growth, if not an enabler for the growth of other sectors. As can be seen in Fig. 1, some innovative products that were developed in this industry have been generally adopted to improve the quality of life of society.

Although the use of technological innovations and developments has been present since the origins of space exploration, in the 1950s. With the emergence of new space technologies about 15 years ago, a substantial rise of space applications took place and, finally, key policy decisions to support the development of a new ecosystem of commercial space actors were fostered. New space encompasses aspects such as small satellites, tight manufacturing processes, multimillion-dollar technology companies investing in space projects and paving the way for more private investment, advances in computing power and digitalization.

[5] Organization for Economic Cooperation and Development [OECD], (2012), OECD Handbook on Measuring the Space Economy. https://www.oecd-ilibrary.org/economics/oecd-handbook-on-measuring-the-space-economy_9789264169166-en.

This has resulted in a reduction in the cost of access to space technologies and has fostered greater space innovation. In just over a decade, "new space players" have had a major impact on the global space sector, providing new disruptive offerings in space manufacturing and operations, launch services, as well as in specific applications such as Earth observation and (satellite) communications. One of the most used tools within this process in recent years is presented by AI.[6]

Therefore, the objectives of this research are to identify the main applications of AI in the space sector, as well as its contribution to the fulfillment of the National Program of Space Activities 2020–2024 of the Mexican Space Agency. To this end, the first stage briefly presents the contributions of the theory of economic change, whose main contribution is the use of technology to improve production processes and generate growth and economic development.

In a second stage, a brief description of what AI is, some of its components and applications is presented. In the third section, a literature review was carried out and the information obtained through the creation of knowledge networks was analyzed. Then, secondary sources of information were consulted to identify technological trends associated with AI that impact the space sector and that can also be used in the Mexican case, especially in industries such as tourism, agro-industrial, manufacturing and construction. Finally, some conclusions are presented.

2 Innovation and Technological Developments to Promote Economic Growth

Economic growth and the explanation of why countries grow and why some produce more than others have been at the heart of economic science since its inception as an academic discipline. In fact, these questions motivated Adam Smith, the father of modern economics, to write his Inquiry into the Nature and Cause of the Wealth of Nations in 1776. However, it was not until the middle of the twentieth century that the first formal models of economic growth were generated from the works of F. Harrod,[7] E. Domar,[8] and especially Robert Solow[9] with his neoclassical model of economic growth. With these models, and mainly because of their explanatory limitations to the real economy, a strong trend began in the economic discipline to study the determinants of growth. Of these determinants, one of the most recognized, in addition to capital and labor, is that of innovation. Authors

[6] Organization for Economic Cooperation and Development [OECD], (2023), Harnessing "new space" for sustainable growth of the space economy.

[7] Harrod, R.F. (1939). An essay in dynamic theory. Economic Journal. Vol 49 (193). Pp 14–33.

[8] Domar. E. (1946). Capital Expansion, Rate of Growth, and Employment. Econometrica Vol. 14, No. 2 (Apr., 1946), pp. 137–147.

[9] Solow, R. (1956). A Contribution to the Theory of Economic Growth. The Quarterly Journal of Economics, Vol. 70, No. 1 (Feb., 1956), pp. 65–94.

such as Schumpeter,[10] Solow himself,[11] Abramovitz,[12] Schmookler,[13] Griliches,[14] Fagerberg[15] and Freeman,[16] recognize innovation as a key factor for development and economic growth. On the other hand, authors such as Freeman,[17] Porter[18] and Nelson,[19] point out that obtaining new and advanced technologies is an important determinant of the competitive position of a country or region, therefore innovation would be the only way for a country to generate, in the long term, a better competitive position and sustainable economic growth.[20]

The last decades have been marked by massive incorporation of new technologies that have redefined not only the innovative, productive, commercial, financial routines, etc., but also the very way of life of a part of the inhabitants around the world. The so-called new information and communication technologies (ICT) are at the heart of this transformation, to which must be added other technologies that are even less advanced in their development, but also have broad effects (biotechnology, new materials, among others).[21] Similarly, the dynamics of competition have visibly shifted towards an increasing weight of technological capabilities as determinants of the economic performance and competitiveness of companies and countries.[22]

---

[10] Schumpeter, J. A. (1939). Business Cycles: A Theore- tical, Historical and Statistical Analysis of the Capitalist Process. Primera Edición, McGraw- Hill, New York.

[11] Solow, R. (1956). A Contribution to the Theory of Economic Growth. The Quarterly Journal of Economics, Vol. 70, No. 1 (Feb., 1956), pp. 65–94.

[12] Abramovitz, M. (1956), Resource and Output Trends in the United States since 1870." National Bureau of Economic Research, Vol. Ocasional Paper 52, pp. 5–22.

[13] Schmookler, J. (1966) Invention and Economic Growth. Harvard University Press, Cambridge, MA. https://doi.org/10.4159/harvard.9780674432833.

[14] Griliches, Z. (1986). Productivity, R&D, and the Basic Research at the Firm Level in the 1970's. American Economic Review, vol. 76, issue 1, 141–54.

[15] Fagerberg, J. (1988). Why growth rates differ. G. Dosi, et al. (Eds.), Technical Change and Economic Theory, Pinter, London (1988), pp. 432–457.

[16] Freeman, R.E. (1994), The Politics of Stakeholder Theory. Business Ethics Quarterly, 4, 409–421. https://doi.org/10.2307/3857340.

[17] Freeman, C., (1987), Technology Policy and Economic Performance: Lessons from Japan. Printer Publishers.

[18] Porter, M., (1990), The Competitive Advantage of Nations. Harvard Business Review. https://hbr.org/1990/03/the-competitive-advantage-of-nations.

[19] Nelson, R.R. (1993), National Innovation Systems: A Comparative Analysis. Oxford University Press, Oxford.

[20] Baumert, T., Buesa, B. M., Gutiérrez R. C. and, Heijs, J., (2016), Innovation and economic growth. Working Paper No. 101. Complutense University of Madrid. https://www.ucm.es/data/cont/media/www/pag-41192/DT101%20-%20Innovación%20y%20Crecimiento%20Económico.pdf.

[21] Bolivar, R. H. and Marronquín, A. I., (2013), Technological innovation as a mechanism to boost economic growth. Regional evidence for Mexico. Count. Adm vol.58 no.3 Mexico City Jul./Sep.

[22] Ibid., 22.

Therefore, the relationship between investment in R&D, technological innovation and economic growth has been studied from two major perspectives: in the area of economic growth (macroeconomics) through theoretical models in which technological innovation is an endogenous factor of growth (evolutionary theory). The other vision contemplates the study of industrial organization through models that aim to explain the behavior of companies in relation to with activities related to innovation (both in its generation and in its adoption) in the market economy.[23]

The evolutionary theory of economic change proposed by Nelson and Winter[24] focuses on the behavior of technological advances in firms that are indicative of economic growth and the adaptations that take place, in line with the transformations of the social and economic environment.[25] According to Pavitt and Soete,[26] Fagerberg[27] and Dosi et al.[28] technological advances are considered one of the main determinants of economic growth; by empirically demonstrating that there is a very close relationship between these two variables.

The central point of these endogenous growth models is that technological innovation is in some way conducive to sustainable economic growth, since there are constant returns to innovation in terms of human capital employed in the R&D sectors.[29]

3 Classification and Description of Artificial Intelligence

The research on AI is descriptive and is based on the literature review. It will provide a brief definition of what AI is, describes some of its components and respective applications are presented. With this information, knowledge networks were generated that allowed us to establish the main uses and applications of AI in the space sector.

Subsequently, this information was used to identify technological trends associated with AI that impact the space sector and that can also be used in the Mexican case, especially in sectors such as agribusiness, manufacturing and construction.

---

[23] Ibid., 13.

[24] Nelson, R. and Winter, S. (1982). An Evolutionary Theory of Economic Change. The Belknap Press of Harvard University Press. Cambridge.

[25] Lovera, M. I., Castro, E., Smith, H., Mujica, M., and Marín, F., (1). Economic evolutionism from the perspective of Nelson and Winter. Multiciencias, 8. https://www.produccioncientificaluz.org/index.php/multiciencias/article/view/16721

[26] Pavitt, K., Soete, L., (1981). International differences in economic growth and the international location of innovation.

[27] Fagerberg, J. (1988). Why growth rates differ. G. Dosi, et al. (Eds.), Technical Change and Economic Theory, Pinter, London (1988), pp. 432–457.

[28] Dosi, G., Freeman, C., Nelson, R., G. Silverberg, and L.G. Soete (Eds.), (1988), Technical Change and Economic Theory, Pinter, London.

[29] Bolívar, R. H., and Marronquín, A. I., (2013), Technological innovation as a mechanism to boost economic growth. Regional evidence for Mexico. Count. Adm vol.58 no.3 Mexico City Jul./Sep.

3.1 General Characteristics of Artificial Intelligence: general characteristics

For a long time, the study of intelligence has been approached from different fields such as psychology, philosophy, pedagogy, to name a few; because it is considered a valuable and decisive attribute to obtain educational and labor achievements and promote the economic development of nations[30] According to Ebbinghaus,[31] intelligence is the ability to adapt to new situations; according to Binet and Simon[32] it refers to formal qualities such as memory, perception, attention and intellect and is characterized by understanding, invention, direction and censorship.

In addition to the above, Rajendra[33] proposes that intelligence is a person's ability to understand the objective world and apply knowledge to solve problems and is made up of different capacities, including perceiving and understanding objective things, acquiring experience and knowledge through learning, understanding and applying that knowledge, analyze and solve problems, reason, make decisions, innovate, create and adapt to complex environments quickly and reasonably as well as predict and understand the development and changes of things.

Therefore, his analysis has not only focused on creating a common concept, but studies have also been carried out to identify how the mind operates and in this sense Fodor,[34] establishes that the faculties of the modular mind "operate employing abstract algorithms like computers".

According to Manod,[35] all living beings, from single-celled to mammals, are "intelligent machines" that are built in the attempt to survive and generate complex variable knowledge mechanisms; all of them have the capacity to process the information necessary to live. Of these last assertions, the idea of associating the functioning of the human brain with a computer where machines try to imitate human cognitive functions.[36]

---

[30] Ardilla, R., (2011), Inteligencia. ¿Qué sabemos y qué nos falta por investigar? Rev. acad. Colomb. science. exact. fis. NAT. vol.35 no.134 Bogotá.

[31] Ebbinghaus, H. (1885). Memory: A contribution to experimental psychology (H. A. Ruger & C. E. Bussenius, trans.), New York: Teachers College, Columbia University.

[32] Binet, A. and & Simon, Th., (1905), Méthodes nouvelles pour le diagnostic du niveau intellectuel des anormaux', L'Année Psychologique, 11, pp. 191–244.

[33] Rajendra, A. (2019), Introduction to Artificial Intelligence. In: Artificial Intelligence for Business. Springer Briefs in Business. Springer, Cham. https://doi.org/10.1007/978-3-319-97436-1_1.

[34] Fodor, J., (1983), The Modularity of Mind. MIT Press.

[35] Manod, J., (1970), Chance and necessity. Barcelona, Barral.

[36] De Val, P. I. (2021), La inteligencia humana y la inteligencia artificial. Anales de Psicologia nº 186:11.

Although the term AI is commonly used today, it is not a new term. It was coined by John McCarthy et al.[37] in 1956 in a conference at Dartmouth College, New Hamspshire and since its origin it has raised the discussion on how to simulate human intelligence through machines.

Therefore, it is established that AI is a branch of computer science that tries to reproduce human cognitive functions such as reasoning, memory, judgment or decision and, subsequently, entrust a part of these faculties, considered as signs of intelligence, to computers.[38]

In this sense, Sosa states that this term does not refer to robots, but to machines that perceive environments and learn by themselves, solve and make decisions autonomously.[39] It is a combination of algorithms, processes and set of systems, responsible for developing programs that emulate human intelligence through the study, assimilation and learning of information from actions carried out by human beings, one of its objectives being the improvement of automated processes.

To understand how AI works, it is necessary to specify concepts such as Machine Learning (ML) and Deep Learning (DL), which are usually used as synonyms, however, each one has its own characteristics. As can be seen in Fig. 2, AI comprises those techniques and procedures that try to provide machines with intelligence. Secondly, machine learning, often used as a synonym for AI, is really a subfield of AI that addresses algorithms capable of predicting based on previous experience. Finally, deep learning is a subset of ML applied to large amounts of data and based mainly on the use of neural networks.

The impact of technologies associated with AI has grown exponentially in recent years, due to the intensive use of mobile devices that allow capturing a large amount of data, as well as access to the internet; For this reason, scientists are paying more and more attention to the functioning and implementation of autonomous learning with the idea of transforming production processes, the economy and life in general.[40]

For this reason, research and practice have led to finding different applications for AI.[41] Some of these applications are summarized in Table 1.

Currently, AI is applied to various activities carried out by humans, having a greater application in robotics, artificial vision, learning techniques and knowledge management. In the productive sector, the inclusion of this technology provides

[37] McCarthy, J., Minsky M., L., Rochester N., and Shannon C., E., (1956), The Dartmouth Summer Research Project on Artificial Intelligence.

[38] Bourcier, D., (2003), Artificial intelligence and the law. Barcelona, Spain.

[39] Sosa, H., (2022), Artificial intelligence and the metaverse in education. https://www.uacj.mx/detalle.html?ID=29155.

[40] Izzo, D., Martens, M., and Pan, B., (2019), A survey on artificial intelligence trends in spacecraft guidance, dynamics and control, Astrodynamics Vol. 3, No. 4, pp. 287–299.

[41] Aguirre, D., Ceja, J. and Pineda, D., (2024). Inteligencia artificial como alternativa de desarrollo de las PyMES mexicanas. International Network of Researchers in Competitiveness XVII Congress 978-607-96203-0-12.

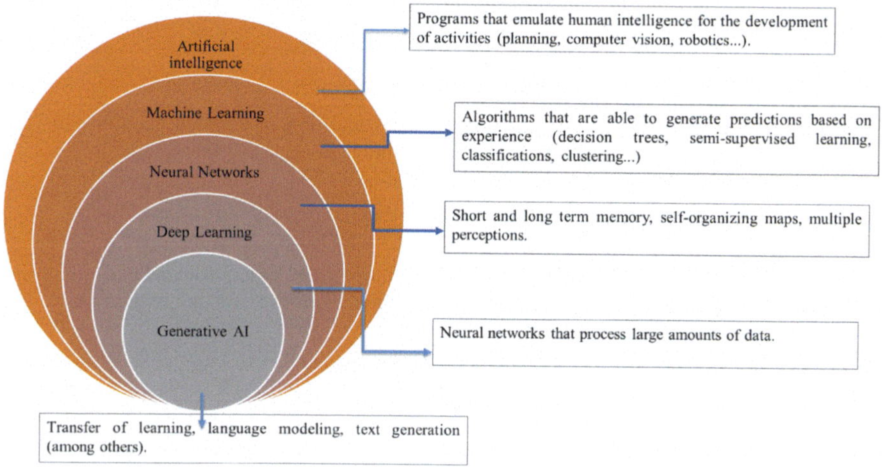

Fig. 2 Diagram of artificial intelligence components. *Source* Prepared by the authors with information obtained from the literature review

Table 1 Applications of artificial intelligence

Application	Description
Visual recognition	Systems capable of recognizing and tracking objects and people in images and videos
Natural language recognition	Systems capable of recognizing, artificially reproducing and deciphering the meaning of spoken language, also includes automatic translation between different languages, as well as automatic answers to questions and the analysis and synthesis of documents
Strategy and Planning	Systems capable of generating optimized strategies to solve highly complex and long-term problems. Examples include automated systems, capable of supporting logic and manufacturing tasks, playing video games, or navigating through physical spaces
Diagnosis and decision support	Systems capable of analysing complex problems and helping to make decisions, for example in the medicine field, in the detection of diseases or in the choice of the most appropriate treatment
Human collaboration	It consists of incorporating intelligent systems as part of human work teams. For example, to respond more quickly to natural disasters, systems have been developed that can analyse aerial views of affected areas to identify where the most support is needed

Source Aguirre, D., Ceja, J. and Pineda, D., (2024), Inteligencia artificial como alternativa de desarrollo de las PyMES mexicanas. International Network of Researchers in Competitiveness XVII Congress 978-607-96203-0-12.

Table 2 Uses and applications of artificial intelligence in the space industry

Uses and applications	Examples
Data analysis	Identify trends, anomalies, strategic planning
Spacecraft operation	Autonomous systems, real-time monitoring, fault diagnosis and detection, communication management, minimizing energy consumption
Predictive maintenance	Predict potential failures, extend the life of parts and components, identify anomalies
Mission planning	Identify and propose new trajectories, efficient use of resources, monitoring of objectives and activities, risk management, contingency plans
Stand-alone systems	Navigation, data analysis and processing, use of robots in maintenance, altitude control and orbit selection, crew health monitoring
Image and signal processing	Planetary surface mapping, climate, atmospheric analysis, detection of cosmic radiation, detection of asteroids and comets, advances in understanding the universe.
Robotics and telerobotics	Exploration on other planets, satellite maintenance, asteroid sampling
Crew support	Personalized recommendations, physical and mental health monitoring, task management, effective communication, training and skill development, sleep quality monitoring and social interaction
Strategies for apace exploration	Planning and prioritization of space missions to meet scientific, economic and social objectives Data analysis

Source Authors' elaboration based on the literature review

automatic methods for decision-making. Manufacturing uses robots that have a great capacity to move and locate objects[42]

According to the Federal Institute of Telecommunications[43] AI will continue to impact the economy of the states and productive systems, which can be observed from the automation of tasks to the creation of new business models. Table 2 summarizes some current and future impacts of uses and applications of AI in the space industries.

Just as the adoption of AI has brought and continues to bring more benefits, there are also some factors to consider in its implementation, among which are:

[42] Ibid., 8.

[43] IFT (Federal Institute of Telecommunications), (2024), Artificial Intelligence in SMEs: Advantages, Risks. https://www.ift.org.mx/transformacion-digital/blog/inteligencia-artificial-en-las-pymes-ventajas-riesgos0#:~:text=Benefits%20of%20the%20IA%20for%20the%20SMEs&text=Impact%20Current%3,efficiency%20and%20reduction%20of%20costs.

(a) Data security: The collection and analysis of large amounts of data poses significant risks in terms of data privacy and security.
(b) Algorithmic bias: If training data is biased, automated decisions can perpetuate or even exacerbate these biases.
(c) Implementation costs: Although the cost of AI technology has decreased, the initial investment and maintenance can be significant, especially for small and medium-sized enterprises from developing countries.

3.2 Artificial Intelligence and the Space Sector

The aerospace industry is defined by Carricaseaux and Frigant[44] as the set of organizations dedicated to the development of productive activities that guarantee boardtruction and design of airplanes, helicopters, launchers, missiles, and satellites, as well as the equipment on which they depend, as well as engines and appliances used on board of aircraft. This sector is aimed at military or civilian customers, starting with airlines and telecommunications network operators. According to the destination of the products and services offered, the industry differentiates between the space and aeronautical part.

This industry is characterized by being one of the areas that generates the most high-tech goods worldwide, therefore, innovation is one of the most relevant activities from the design to the operation of aircraft. It is also associated with national security issues due to its close relationship with armaments, communications and defense.[45]

Among the particularities of the aeronautical industry are:(i) need for large investments, (ii) return on investment and very long production cycles, (iii) certifications attached to the requirements of the industry and (iv) high technological level in design and manufacturing; (v) strong reliance on innovation and research and development (R&D) activities[46]; (vi) high-cost and high-risk programs, (vii) production in short series and with high added value.[47]

To identify the relationship between the space industry and AI, a search was carried out in specialized databases using the keywords “space industry” AND “AI”. The results of this study were 79 documents, which were analyzed through

---

[44] Carrincazeaux, C. & and Frigant, V., (2007), The Internationalization of the French Aerospace Industry: To What Extent Were the 1990s a Break with the Past? Competition and Change, 11(3), pp. 261–285.

[45] Gereffi, G. & and Sturgeon, T., (2013)., Global Value Chain-Oriented Industrial Policy: The Role of Emerging Economies. Elms, Deborah & and Low, Patrick (Eds.), Global Value Chains in a Changing World. Geneva: World Trade Organization, pp. 329–360.

[46] Niosi, J. and Zhegu, M., (2005), Aerospace Clusters: Local or Global Knowledge Spillovers? Industry and Innovation, 12:1, pp. 5–29, https://doi.org/10.1080/1366271042000339049.

[47] Hernández, J., (2015), Mexican companies in the global value chain of the aeronautical industry. Mexico: FLACSO.

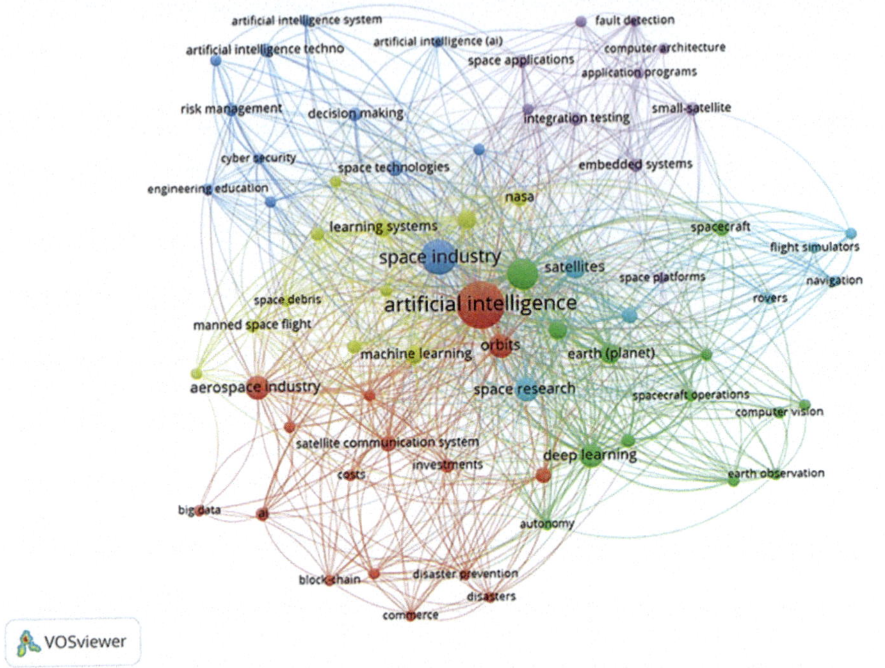

Fig. 3 Analysis by Co-occurrence of Keywords "Space Industry" AND "Artificial Intelligence". *Source* Authors' elaboration with information obtained from Scopus and processed with VoSviewer software.

the creation of a knowledge network based on co-occurrence by keywords (see Fig. 3).

As can be seen in Fig. 3, the main uses given to AI in the space industry focus on the development of communication systems, autonomy, application in satellites, cybersecurity and risk management. Table 2 details these applications in the space sector.

3.3 Space Sector in Mexico

Space is the scene of great challenges and opportunities for humanity. Over the past few decades, we have witnessed impressive advances in the exploration and use of space, such as the operation of exploration probes on other solar system bodies such as the Moon and Mars, the employment of reusable launchers, and the development of low-orbit constellations to provide high-speed communications services to the entire planet.[48]

According to the National Program of Space Activities 2020–2024 of the Mexican Space Agency,[49] Mexico has several strengths in the space field, which has been acquired since the launch of the first generation of Mexican satellites in 1985 (Morelos 1 and 2), and the first Mexican astronaut whit the scientific experiments he carried out in space. New generations of spacecraft followed with the Solidaridad 1 and 2 satellites, then the Quetzal, the Satmex 5 and currently the Eutelsat 6, 7, 8, 9 satellites and the Bicentenario and Morelos III satellites.

However, it is also recognized that Mexico is primarily a consumer of space technologies and applications. The reasons for that is that the development of national space capability has been hampered by the capital-intensive nature of the space sector and the lack of coordination between national research, Government, and private sectors to make an overall collective effort. Therefore, it is proposed that if these adversities are overcome, not only will there be an important advance in a worldwide strategic sector, but it will also benefit the development of the country in the economic, political, environmental and social spheres.[50]

Therefore, the following objectives are proposed for Mexico and its National Programmed of Space Activities 2020–2024:

1. Identify perspectives and promote the development of spatial infrastructure for telecommunications, navigation, global positioning and their applications, which favor digital transformation and the offer of services to contribute to well-being, social inclusion and economic development.
2. Promote the development of a comprehensive national Earth observation program that meets the information needs of the pentahelix for the benefit of the population.
3. Increase national capacities and promote cooperation in science and technology, in space exploration for the scientific and technological strengthening.

In this sense, adopting some of the technologies and processes associated with AI could contribute to the achievement of these objectives of the National Programme of Space Activities (see Table 3).

Meeting the aforementioned objectives is of utmost importance to advance in the growth of the space knowledge, education, science and industry. This will generate and obtain benefits from the technologies developed directly and indirectly.

[48] Duarte, C., (2024), How to participate in Mexico's space development? Stories you should know. https://haciaelespacio.aem.gob.mx/revistadigital/articul.php?interior=1554.
[49] AEM, (2020), National Program of Space Activities 2020-2024. https://www.gob.mx/cms/uploads/attachment/file/585644/Programa_Nacional_de_Actividades_Espaciales_2020-2024.pdf.
[50] Ibid.

Table 3 Contribution of artificial intelligence to the achievement of the objectives of the national programme of space activities 2020–2024

Objective	Uses and applications of AI	Description
1. Identify perspectives and promote the development of spatial infrastructure for telecommunications, navigation, global positioning and their applications, which favor digital transformation and the offer of services to contribute to well-being, social inclusion and economic development	Data analysis Spacecraft operation Image and signal processing	Using AI to collect and process large amounts of data will allow the implementation of predictive systems that help correct and timely decision-making
2. Promote the development of a comprehensive national Earth observation program that meets the information needs of the pentahelix for the benefit of the population	Data analysis Stand-alone systems Robotics and telerobotics Image and signal processing	The data collected can be used to anticipate latent risks associated with climate change, infrastructure planning, health and education
3. To increase national capacities and promote cooperation in science and technology, in space exploration for the scientific and technological strengthening	Data analysis	The information collected can be used for the development of Research and Development and Innovation (R&D+I) projects, in addition to being used in the training of specialized human capital in the space industry

Source Authors' elaboration with information from Tables 1 and 2.

Directly, the population uses information and telecommunications services associated with space. Planet Earth monitoring applications are also used that can warn about risks associated with climate change.[51]

Indirectly, there are benefits in computer technologies, consumer products, energy, medicine, industrial productivity, manufacturing technology, microelectronics and transportation, among others. All this contributes to the generation of jobs and the creation of technology-based companies. Space technology is also a great opportunity to strengthen international cooperation by carrying out joint and international projects, exchanging information and experts.[52]

According to Duarte, space is a multidisciplinary field that requires knowledge and skills in basic sciences, engineering, technology, business, law, education, and

[51] AEM, (2020), National Program of Space Activities 2020-2024. https://www.gob.mx/cms/uploads/attachment/file/585644/Programa_Nacional_de_Actividades_Espaciales_2020-2024.pdf.

[52] Ibid.

communication, among many others, and Mexico has a large amount of human capital to create innovative solutions to develop national ventures.[53]

3.4 Applications of Artificial Intelligence in the Space Sector

Among the sectors that contribute the most to the Gross Domestic Product (GDP) at national level are manufacturing, commerce, information and communication technologies, agri-food, construction, and tourism.[54] All of them, with the potential to be driven by the use of AI. Table 4 summarizes some of the current and near future impacts that the use of AI will have on production systems in general.

In addition to the above, technologies associated with the space sector can be applied specifically to accelerate and strengthen the sectors that contribute the most to GDP at the national level. In general, this will allow companies to be more efficient and effective, increase their productivity as well as reduce delivery times and risks. Table 5 provides an exemplary overview of applications of space technologies in the growth of productive sectors in Mexico.

As can be seen in Table 5, the space industry and the technologies associated with it are a source of social and economic benefits for nations that invest in it due to the impact they have on productive processes, innovation, competitiveness and growth of economies, in addition to generating qualified jobs. The above not only occurs directly, it also happens in economic sectors that indirectly support the space sector and, at the same time, benefit from it. In this sense, the use of AI has proven its importance and effectiveness to find solutions that improve productivity while reducing risks specific to each sector and those that currently present themselves in a disruptive manner, thus allowing the right decisions to be made with their respective benefits.

Artificial intelligence plays a crucial role in achieving the aims of the National Space Activities Program 2020–2024 of the Mexican Space Agency. The artificial intelligence should be used to enhance decision- making through the analysis of the large of data sets, optimize spacecraft operations, process images and signals and predict risks related to infrastructure or climate change. Moreover, artificial intelligence can help develop autonomous programs and robotic systems that improve space observations and exploration.

In this way, artificial intelligence can be the key enabler in meeting these goals by streamlining data collection and analysis, enhancing planetary observation and supporting the management of natural resources.

[53] Duarte, C., (2024), How to participate in Mexico's space development? Stories you should know. https://haciaelespacio.aem.gob.mx/revistadigital/articul.php?interior=1554.

[54] Instituto Nacional de Estadística y Geografía [INEGI], (2024), Economía y Sectores Productivos. https://www.inegi.org.mx/temas/pib/.

Table 4 Impacts of artificial intelligence on production systems

Current impact	Near future impact
Automation and efficiency: AI has enabled the automation of repetitive and routine tasks in industries such as finance, customer service, and operations. This has led to greater efficiency and cost reduction	**Deeper integration in the value chain**: AI will be further integrated into all stages of the value chain, from research and development to production and distribution
Personalization: E-commerce, streaming, and advertising companies use AI to analyze user behavior and offer personalized recommendations, improving the customer experience and increasing sales	**Expansion into traditional sectors**: Sectors that have been slower to adopt AI, such as agriculture and manufacturing, will begin to see more significant transformations driven by the technology
Large-scale data analysis: AI, especially machine learning, has revolutionized the ability of companies to analyze large data sets, obtaining valuable insights for decision-making	**Ethics and regulation**: As AI becomes an integral part of the digital economy, debates and regulations will emerge about ethics, privacy, bias, and transparency
New Business Models: AI has enabled the creation of new business models, such as virtual assistant platforms, chatbots, and solutions based on predictive analytics	**Employment and training**: While AI will create new jobs and opportunities, it will also displace certain roles. This will require a reconfiguration in job education and training to adapt to an evolving labor market
Finance and banking: AI are used for fraud detection, risk management, algorithmic trading, and automated financial advice	**Explanatory AI**: As AI-based decisions become more critical, the demand for AI models that not only predict but also explain their decisions in understandable ways will grow
Improved customer service: AI-powered chatbots and virtual assistants are transforming customer service, offering quick and accurate responses to user queries.	**Human-machine interaction**: Interfaces and interaction are expected to become more intuitive and natural, possibly through advances in virtual reality, augmented reality, and emotional response systems
Cybersecurity: Prospective research explore distinct legal framework that successfully govern implementations of AI in cybersecurity across different nations	**AI in cybersecurity:** AI will play a crucial role in proactive threat detection and incident response, but challenges will also arise in terms of AI-powered attacks

Source Prepared by the author with information obtained from IFT (2024).

4 Conclusions

The adoption of AI has the potential to contribute to the resolution of global problems such as climate change, the early detection of natural phenomena, disaster management, as well as impact the improvement of people's quality of life in specific issues such as support for education, access to health systems and reduction of the food crisis.

Among the main applications of AI in the space sector are the development of communication systems, autonomy, application in satellites, cybersecurity and risk management. It was identified that these can contribute to the fulfillment of the

Table 5 Application of space technologies in the growth of productive sectors in Mexico

Sector	Momentum through Space Technologies[55]
Manufactury	Supply chain and transportation will benefit from more efficient and cost-effective logistics. In addition to having boots that allow the automation and optimization of processes, reduce waste and risks
Commerce	It is used to streamline decision-making processes, improve data analysis, and optimize supply chain management. It can also predict market trends, automate customer service and improve inventory management, making internationalization more efficient and reliable
Food	Space technologies can be vital for agricultural innovation, modern agriculture, and precision agriculture. They provide weather and drought forecasting services to farmers, pastoralists and fishers to promote sustainable agricultural development, increase agricultural productivity and contribute to food security, as well as produce fertilization maps and applications, and disseminate satellite imagery and analysis data to farmers via the Internet In addition, the adoption of some applications can help monitor crops from space using public sources of satellite data and algorithms on land use and land area
Construction	Increase productivity, optimising resources and improving compliance with deadlines and the quality of projects and works. In the case of smart buildings, they make it possible to effectively meet people's needs and improve accessibility to buildings. In addition, it reduces costs and space problems, helping to improve the adaptability of buildings. It also helps improve building management by automating all building systems

Source Prepared by the author with information obtained from UNCTAD[56] (2020).

National Program of Space Activities 2020–2024 of the Mexican Space Agency by collecting and processing large amounts of data that allows the implementation of predictive systems that help correct and timely decision-making.

The data collected can be used to anticipate latent risks associated with climate change, infrastructure planning, health and education. In addition, it can be used in the development of Research and Development and Innovation (R&D+I) projects, as well as in the training of human capital specialized in the space industry.

In view of the above, the use of AI stands out as a tool with multiple areas of action and benefits that go beyond technical issues, since its application in various economic sectors, specifically the space sector, has a wide range of benefits. Possibilities that, if used correctly, will allow finding new solutions to global, regional

[55] Theterm "spacetechnologies" usuallyincludessatelliteEarthobservation, satellitecommunications and satellitepositioning. Spacesciences and technologies are central to technologiessuch as thoseused in weatherforecasting, remotesensing, global positioningsystems, and satellite television and communications systems, as well as to scientific fields such as astronomy and Earth sciences. (UNCTAD, 2020).

[56] UNCTAD, (2020), Contribution of space technologies to sustainable development and advantages of international collaboration in research in this field. https://unctad.org/system/files/official-document/ecn162020d3_es.pdf.

and national problems and challenges, thus decision makers will have a solid foundation to build an institutional matrix that responds to the major questions that currently affect the international and national situation.

Artificial intelligence helps to meet the goals proposed by the AEM; thus, playing a role in the creation of autonomous systems for space exploration and the improvement of planetary observation and resource management.

Castillo Malagón Romina B.A. in International Relations, International Relations, master's in international business administration and PhD in Management Sciences from the National Autonomous University of Mexico. During 2017, she completed a research stay at the Faculty of Economics and Business at the University of Salamanca, Spain. Since 2011, she has worked at the National Polytechnic Institute mainly in the development of projects linked to the productive and governmental sectors, such as training and updating programs for agencies like the training and updating programs for agencies such as the Federal Electricity Commission (CFE), National Water Commission (CONAGUA), Airports and Auxiliary Services (ASA), Metro Collective Transport System and the Institute of Security and Social Services for State Workers (ISSSTE). She has participated in the development of technology-based projects funded by the National Council of Science and Technology (CONACYT) and as a speaker at national and international conferences, among her lines of research are innovation, entrepreneurship, sustainability, gender equity and technological development in the aviation industry. She has held public positions as head of the Department of Development of Technological Offering at the Technological Development Unit of the IPN and Head of Liaison and Technological Development at the Technological Institute of Gustavo A. Madero II. From 2019 to 2020, she worked as an analyst at the Directorate of Prospective and Technological Intelligence, TechnóPoli of the IPN. Technological Intelligence, TecnóPoli of the IPN.

Farah-Simón Lisette Completed her undergraduate studies at the Faculty of Engineering at UNAM (National Autonomous University of Mexico) where she graduated as an Industrial Engineer; she obtained a master's degree in mechanical engineering from the same faculty and received the degree of Doctor of Administration Science, with honours, from the FCA (Faculty of Accounting and Administration -UNAM). She has participated in several R&D projects, among which her participation stands out: in the Macro Project "Administration and Sustainability" of the FCA with the project management of technology with a sustainable perspective in the space industry in Mexico; in the Space Science and Technology Network (Thematic Network) and in the University Space Program, with a project on space sustainability; as well as the research she carried out on the design of an adjustable socket for lower limb prostheses; which has been protected through an invention patent. She is a member of the National System of Researchers of Conacyt (National Council of Science and Technology).

Buenrostro Aguilar Hugo Javier Degree in International Relations (Faculty of Political and Social Sciences (FCPyS-UNAM); master's in international business administration (Faculty of Accounting and Administration, FCA-UNAM); Doctorate in Administration Science (Faculty of Accounting and Administration, FCA-UNAM). Full-time professor attached to the Research Division of the Faculty of Accounting and Administration (FCA) of the UNAM (National Autonomous University of Mexico). Lecturer and guest speaker by the Centre for Social Studies and Public Opinion, Chamber of Deputies; the Strategic Research Institute of the Mexican Navy and the Centre for Naval Superior Studies, Mexican Navy Secretariat; and, by the Superior School of War of the Secretariat of National Defense. Member of the Mexican Association of International Studies; the Administration and Sustainability Macroproject of the FCA-UNAM; and, from the University Seminar on Social Entrepreneurship, Sustainable Administration and Comprehensive Training at Middle and Higher Levels at the National Autonomous University of Mexico (SUESA). He is a member of the National System of Researchers of Conacyt (National Council of Science and Technology).

Applications and Implementation of Design Methodologies in Mexican and Latinamerican Aerospace Projects

Pablo Alejandro Arizpe Carreón,
Jhonatan Fernando Eulopa Hernandez,
Diego Alfredo Padilla Pérez, Carlos Couder Castañeda,
and Edgar Bibiano Valente

Abstract

Increased investment, technological advancements, brand-new education initiatives, and strategic partnerships have all contributed to the rapid growth of the aerospace industry in Mexico and Latin America. As professionals in this field, your role is crucial in this growth. As the region continues to develop its aerospace capabilities, your expertise in implementing effective design methodologies becomes even more vital to the success of aerospace projects. These methodologies, including systems engineering, concurrent engineering, and agile development, are essential for managing the complexity of aerospace systems and ensuring that projects meet stringent safety, performance, and cost requirements. In the context of Mexican and Latin American aerospace projects, developing region-specific design methodologies will enhance the region's ability to compete globally. This chapter will explore the various applications of newly developed design methodologies in the Mexican aerospace sector,

P. A. A. Carreón (✉) · J. F. E. Hernandez · D. A. P. Pérez · C. C. Castañeda · E. B. Valente
Aerospace Development Center, IPN, Mexico City, Mexico
e-mail: parizpec@ipn.mx

J. F. E. Hernandez
e-mail: jeulopah@ipn.mx

D. A. P. Pérez
e-mail: dpadillap@ipn.mx

C. C. Castañeda
e-mail: ccouder@ipn.mx

E. B. Valente
e-mail: ebibanov@ipn.mx

J. A. Ferrer Pérez and A. Froehlich (eds.), *Space Fostering Mexican Society*, Southern Space Studies, https://doi.org/10.1007/978-3-031-92617-4_4

emphasizing their impact on project outcomes, collaboration, and technological advancement in the region.

1 Introduction

The NASA Systems Engineering Handbook[1] is a widely respected resource that has become an essential reference for researchers and students in the Mexican Aerospace Sector. The Handbook provides Mexican researchers and students a framework to align their work with international best practices. Nevertheless, Mexican researchers and students may encounter challenges when applying international methodologies to local projects with different regulatory, economic, or cultural contexts. The systems engineering processes described in the NASA Handbook are designed for large-scale, high-budget projects typical of NASA missions. The rigor and complexity of these processes may be overwhelming or unnecessarily burdensome for smaller or less resource-intensive projects, which are common in the Mexican aerospace and educational sectors.

As Mexico's aerospace industry expands, the need for increased educational programs and robust and effective systems engineering practices has grown. International design methodologies offer a comprehensive framework for managing the life cycle of complex aerospace systems, from conception to design, implementation, testing, operation, and out-phasing. Their adoption by Mexican researchers and students underscores the importance of aligning with international standards and methodologies in a field where accuracy and reliability are paramount. This alignment is particularly advantageous when collaborating on multinational projects or seeking integration into the global aerospace supply chain.

Adhering to a rigorous space methodology ensures that students and researchers develop a comprehensive understanding of the entire systems engineering process, which is essential for managing the complexity of aerospace projects. For instance, NASA's Handbook is grounded in decades of experience managing some of history's most challenging and high-risk aerospace projects. By applying these proven methodologies from highly technologically advanced countries in the aerospace sector, Mexican researchers and students can leverage the lessons learned and avoid common pitfalls.

In Mexico and several Latin American countries, implementing the full scope of systems engineering practices outlined in NASA's Handbook often requires significant resources, including time, personnel, and specialized tools. When resources are limited, this can greatly disadvantage students and smaller research teams.

[1] Robert Shishko, Robert Aster, Robert Chamberlain, Patrick Mcduffee, Les Pieniazek, Tom Rowell, Beth Bain, Renee Cox, Harold Mooz and Lou Polaski, NASA systems engineering handbook, http://hdl.handle.net/2060/19960002194, 1995.

Developing and implementing an indigenous and entirely Mexican methodology will be a valuable tool for Mexican researchers and students, providing a robust framework that aligns with global standards. However, careful analysis is required to adapt this methodology to Mexican aerospace projects' specific context and constraints.

Several educational Mexican institutions, including the National Polytechnic Institute (IPN) and the National Autonomous University of Mexico (UNAM) have had notable participation and have been academic leaders in projects related to aerospace technology. Historically, IPN and UNAM have worked in developing technology, including creating artificial satellites and components for experimental validation, and have collaborated with international space agencies and national and foreign universities on technical projects. These collaborations allow the exchange of knowledge and technologies. For the research in this chapter the findings and respective outcomes of the IPN will be discussed in the following chapters. The IPN, with its extensive experience in engineering and technological project design and execution, is well-positioned to develop this space methodology. Its strong foundation in practical and applied approaches will be instrumental in shaping this innovative framework. However, the successful implementation of this methodology will require the collaborative efforts of various institutions, including UNAM, to extend its impact beyond the scope of the IPN.

2 Methodology Implementation in Mexico

According to the Strategic Space Agenda[2] by the Mexican Council of Aerospace Education (COMEA), the central and north regions (Fig. 1) possess the highest aeronautical capabilities in Mexico, with particular strengths in conventional and non-conventional machining and welding. Within the central region, Querétaro has the most diverse capabilities. In contrast, Baja California Norte leads in the northwestern region, demonstrating robust development across most sub-industries. Notably, Nuevo León is recognized as the third state with the most advanced capabilities at the national level. The southern region is the top area in the country for the maintenance sub-industry, whereas the central region dominates in engineering and design capabilities.

The same Strategic Space Agenda mentions a significant growth opportunity in the space sector, which includes research and development of technologies and sustainable energy sources. However, it does not play an essential role in the global space context: according to the 'National Space Activities Program 2020-2024,' the annual budget of the Mexican Space Agency (AEM) since its inception has never exceeded USD 3 million, compared to USD 18,500 million in countries like the

[2] COMEA, Strategic Space Agenda, https://comea.org.mx/Paginas/Agenda_estrategica.html, 2022.

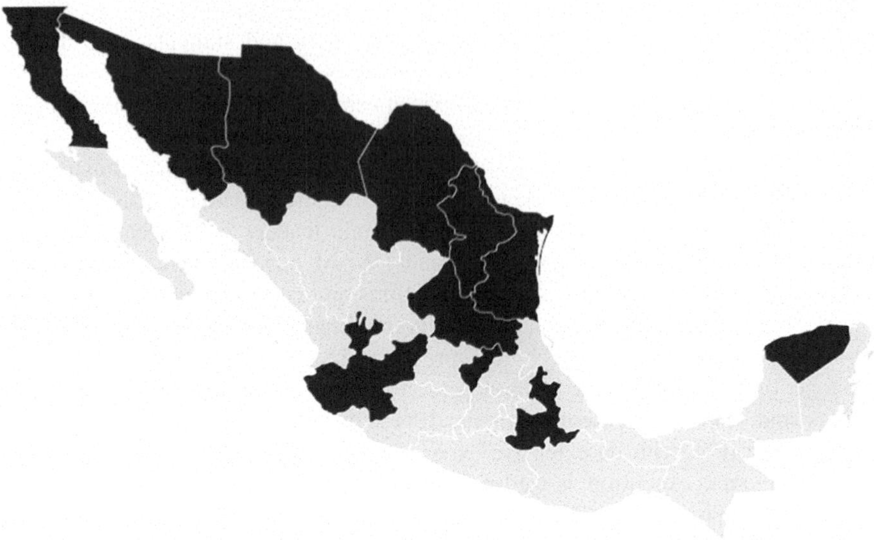

Fig. 1 Aerospace industry in Mexico cutting-edge production and innovation[3]

United States or USD 5600 million in Russia. Argentina and Brazil allocate USD 100 million annually to develop this sector in Latin America.

Considering the last paragraph, it is essential to implement a methodology that aligns with Mexico's indigenous standards and requirements while maintaining overall quality in design, validation, manufacturing, assembly, monitoring, and mission closure.

In the last two years, the IPN has been developing a space methodology[4] that leverages the advantages of the NASA Systems Engineering Handbook while addressing specific challenges faced in Mexico. This methodology assists various other Latin American countries by providing a comprehensive guide on space missions in Spanish language. This space methodology aims to establish a foundational manual for developing space missions, emphasizing the procedures and methods for executing aerospace technology projects. It offers a comprehensive framework to manage the complexity of space missions, encompassing all aspects from requirement definition to implementation, testing, operations, and phasing-out (mission closure).

The IPN space methodology outlines a clear life cycle for aerospace projects, with well-defined phases and decision points. This facilitates effective planning, execution, and monitoring of project progress.

[3] American Industries, Aerospace, https://www.americanindustriesgroup.com/aerospace/, 2024.

[4] IPN website, https://ipn.mx/imageninstitucional/comunicados/ver-comunicado.html?y=2024&n=45&t=6, 2024 (all websites cited in this publication were last accessed and verified on 12 December 2024).

The IPN's space methodology encompasses several key characteristics:

- Language and Collaboration: It is established in Spanish language, facilitating collaboration and coordination among multidisciplinary teams of Latin American origin;
- Usability: It is designed to be straightforward to understand and apply, ensuring ease of use across various stakeholders;
- Risk Management: It supports early identification and mitigation of potential risks, thereby reducing costs and increasing the likelihood of successful mission planning;
- Flexibility: It is adaptable, allowing designers, managers, administrators, chief engineers, and project directors to implement it with enhanced rigor, applying more stringent quality control and supervision techniques than those outlined in the document;
- Phase Structure: It comprises seven phases: conceptual studies, optimization, design, manufacturing, assembly, integration and launch, operation and monitoring, and mission closure. Each phase is subdivided into actions with clear objectives, further broken down into tasks and deliverables;
- Action Sequence: Actions within each phase can be executed in parallel (simultaneous actions) or sequentially (independent actions);
- Conceptual Framework: It addresses the design question "Why do it?", with actions answering "What to do?" and tasks detailing "How to do it?".
- Review and Decision Processes: It incorporates rigorous review and decision. Making processes, with reviews conducted by external peers and decisions made by the project team and responsible authorities;
- Experience Integration: It considers team members' experience and applies to any level of expertise;
- Interdisciplinary Support: It substantially supports various interdisciplinary fields such as geography, urban planning, environmental studies, social sciences, and business;
- Broad Applicability: It applies to all space missions and designs related to space techniques and technologies, regardless of the program's type and size;
- Comprehensive Framework: It offers a comprehensive framework for addressing the complexity of space missions, covering all aspects from requirement definition to implementation, testing, operations, and mission completion, without conflicting with other international methodologies.

3 Methodology Validation

Scientific research is a complex process that integrates various theoretical, methodological, and technical aspects. Before execution, meticulous planning is required to define objectives, establish a reference framework, design activities, and specify necessary resources. Each research endeavor is structured as a project involving the conceptualization, outlining, and organization of a plan or the means for its execution.

Developing an appropriate national indigenous methodology entail aligning resources with objectives and determining the methods used to approach the subject of study. This methodology systematically considers the conditions under which the research will be conducted, ensuring thorough preparation. The procedures of this methodology consist of a series of algorithms and didactic tools designed to guide the researcher in defining and structuring an aerospace research project.

These methodological procedures employ various methods and sequenced approaches to help the researcher comprehend the problem at hand, explore the potential solutions proposed by the author, and assess the conclusions that outline the advantages and disadvantages of the project, ultimately determining whether it effectively addresses the problem.

According to the official website of IPN[5], the concept of integrating and structuring a technical education system emerged in 1932. This initiative featured prominent figures such as lawyer Narciso Bassols and engineers as Luis Enrique Erro and Carlos Vallejo Márquez. The project was realized in 1936 with the support of Juan de Dios Bátiz, then Senator of the Republic, and General Lázaro Cárdenas del Río, Constitutional President of the United Mexican States.

The establishment of a degree of Aeronautical Engineering was approved at the Higher School of Mechanical and Electrical Engineering (ESIME) in 1937. For over seven decades, this program was the sole degree in Mexico dedicated to training professionals in disciplines such as maintenance, operations, air transport, and design in the academic formal aerospace field.

The IPN has played a significant role and has been a leader in aerospace technology projects, collaborating with[6]:

- Regulatory entities
- Space agencies and related organizations
- Federal Government bodies
- Civil associations and trade organizations
- Standardization and certification institutions
- National and international academic institutions.

Within the aerospace projects, the IPN has developed the several projects following precise methodologies in the aerospace field[7]:

- Artificial satellites Painani I and II;

[5] IPN Website, Historical archive, https://www.ipn.mx/decanato/archivo-historio/, 2024.
[6] IPN Website, Academic engagement, https://ipn.mx/cda/apoyo-academico/vinculacion-academica.html, 2024.
[7] IPN Website, Technological Development, https://ipn.mx/cda/desarrollo-tecnologico/desarrollo-tecnologia-aeroespacial.html, 2024.

- Artificial satellites Satex I and II;
- Educational satellites;
- Undergraduate and graduate theses directed by highly qualified specialists;
- Experimental Rockets;
- Design of UAV;
- Space exploration vehicles;
- Remote Control (RC) aircraft;
- CANSAT satellites;
- Ground station development;
- Mexican Space module EMIDSS I-VI[8];
- Training of specialists abroad who contribute their experience to the IPN;
- Construction of functional aircraft;
- Development of specialized laboratories and clean rooms;
- Implementation of augmented and virtual reality in aerospace technology;
- Validation of aerospace instrumentation and equipment;
- Among others.

The IPN has effectively implemented advanced research methodologies across the spectrum of the aerospace field. Supported by a robust academic foundation and substantial infrastructure, the IPN has demonstrated its capability to develop and apply rigorous methodologies to real projects. This enables the institution to tackle complex challenges, drive innovation, and make significant contributions to the advancement of scientific knowledge, particularly in critical areas such as engineering, technology, and applied sciences.

4 IPN Space Methodology

Multiple units within the IPN are involved in developing projects and experimental prototypes, including satellites, rockets, unmanned aerial vehicles, ground stations, and terrestrial space exploration vehicles.

Space methodology can be defined as a comprehensive guide comprising a series of tools and techniques used to conceptualize, analyze, design, and validate space missions. This methodology encompasses various types of analyses applicable during the development of a space mission:

- Descriptive Analysis: Provides a detailed account of space technology operations and explains performance across different operational phases of cosmic devices;
- Qualitative Analysis: Identifies patterns and relationships within data, enabling the inference of various solutions based on expertise and experience;

[8] Diana Saavedra, Prueba la NASA sistema satelital de UNAM e IPN, Experimentación espacial, Gaceta UNAM 5218, pp. 1–3, http://www.acervo.gaceta.unam.mx/index.php/gum10/article/view/92488, 2010-2019.

- Quantitative Analysis: Relies on measurable and verifiable data to ensure the reliability and accuracy of the space mission;
- Inferential Analysis: Conducts statistical tests to identify predictive factors influencing the mission's success.

This methodology applies to all space missions and designs related to space technology and engineering, irrespective of the type or scale of the program. It is important to note that this methodology serves as a guide, providing a series of recommendations that should be followed to ensure the achievement of mission objectives. Each designer, administrator, chief constructor, and project manager can rigorously enforce this methodology by implementing stricter quality control and oversight measures than recommended in this document.

The IPN methodology is employed to guide the development of a space mission across various projects, collectively called "The Life Cycle of a Space Project."[9] The project development process is divided into distinct phases, organized into specific actions, and separated by decision-making milestones. This process addresses a project's conceptualization and validation. During the conceptualization stage, the phases are designated using the nomenclature Alpha, Bravo, and Charlie. In the validation stage, the phases are labeled Delta, Echo, Foxtrot, and Golf. The primary objective of this structured approach is to facilitate effective monitoring and control throughout the project's development, initiation, and end-of-life stages.

The methodology addresses vital aspects of successful project management, emphasizing the importance of clear definition and precise planning. It considers the following components:

- System Planning: Defines and develops a system or space mission that meets stakeholders' requirements and needs;
- Product Development: Involves creating a physical and functional product, verifying and validating it to ensure it meets the requirements and expectations of all involved parties;
- Management and Operation: Encompasses the planning, monitoring, controlling, and directing technical activities necessary for the system's development and management;
- Program Control and Stabilization: Involves detailed project planning with rigorous control measures to ensure time, cost, and performance success.

The methodology also incorporates risk analysis and management (Fig. 2), which is divided into the following components:

- Configuration and Interfaces: Assessment of system configuration and interface integration to identify potential risks;

[9] NASA, SEH 3.0 NASA Program/Project Life Cycle, https://www.nasa.gov/reference/3-0-nasa-program-project-life-cycle/, 2024.

Fig. 2 IPN methodology: risk analysis and management

- Control and Monitoring: Evaluation of control mechanisms and monitoring procedures to manage risk factors effectively;
- Budget and Schedule: Analysis of financial and scheduling aspects to mitigate cost and time constraints risks;
- Product Transition: Management of risks associated with the transition of the product from development to operational deployment;
- Review and Performance: Ongoing review of performance metrics to address any emerging risks and ensure project success.

Project requirements encompass both the classification and assignment of design elements as well as the limitations of the product. Among the requirements outlined in this methodology, the following are particularly notable:

- Functional Requirements: Specify the functions that need to be performed by the system or product;

- Performance Requirements: Define the criteria for how well the functions must be executed;
- Linking Requirements: Detail the degree of interaction between different products or system components;
- Transversal Requirements: Include aspects of the environment, safety, human factors, and standards derived from industry know-how.

Reviews are formal evaluations conducted at specific stages throughout a project's life cycle to ensure that space mission requirements[10] are met, identify and manage unforeseen risks, and validate project designs and plans. These reviews serve as critical control points, ensuring that the project is sufficiently prepared to progress to the next action or phase.

As previously mentioned, this methodology features a distinct characteristic where each phase includes corresponding actions that can be executed either simultaneously (SA-Simultaneous Actions or independently (IA-Independent Actions). Actions labeled as SA can be developed in parallel with other actions sharing the same letter color within the phase, provided they are not in black ink. This does not imply that a subsequent action can be completed before the preceding action; such sequencing is determined by the project manager. Conversely, actions classified as IA must be completed before proceeding to the subsequent action in the planned sequence.

The applied IPN methodology is structured into seven main phases, each defining a specific objective of the project (Fig. 3).

The implementation of this methodology is centered on what the space program manager and project collaborators need to understand and perform to accomplish a space mission. A necessary individually tailored methodology guidance leads in a straightforward and easy-to-understand manner, akin to an instructional manual; however, the scientific and regulatory rigor level is contingent upon the executing team. The actual project may require more specific and detailed actions than those outlined in this document, or, depending on the mission type, it may be more limited.

The number of activities may be reduced at the discretion of the space program manager for student or academic missions. This methodology encompasses seven main phases, each with various activities designed to assist academics in achieving their objectives (Fig. 3).

This document also addresses the analysis and supporting work for decision-making, including cost, schedule, and confidence level evaluations. It suggests conducting activities either sequentially or in parallel to optimize the program manager's time management. Nonetheless, this does not preclude the necessity of collaboration among those responsible for parallel activities.

---

[10] James Wertz, David Everett and Jeffery Puschell, *Space Mission Engineering: The New SMAD*. [Online]. Available: http://ci.nii.ac.jp/ncid/BB0899400X. 2011.

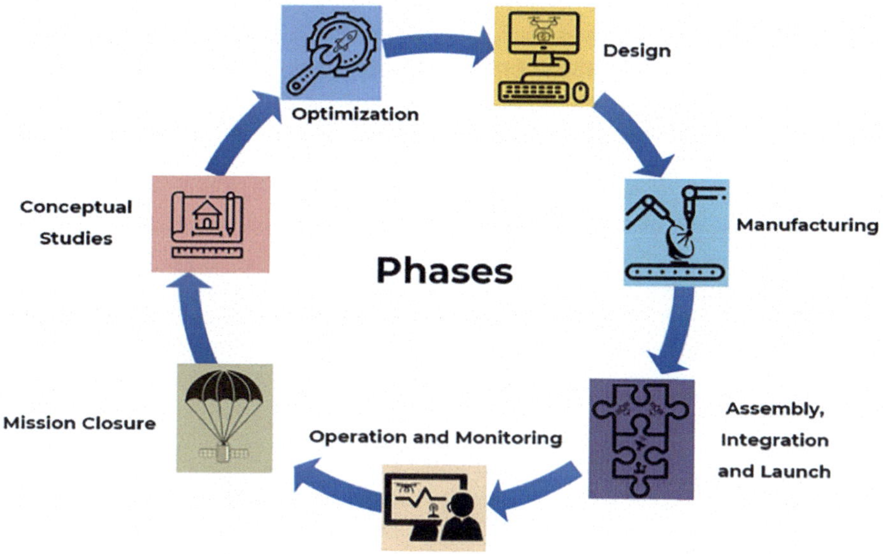

Fig. 3 Applied IPN methodology in seven main phases

5 Conclusions

In general, the methodology emphasizes the importance of clear definition, precise planning, and risk management. It integrates various analyses—descriptive, qualitative, quantitative, and inferential—to support the project's success. By offering a systematic approach to defining functional, performance, linking, and transversal requirements, the methodology ensures thorough preparation and effective execution.

The IPN space methodology described for space mission development is a comprehensive framework designed to guide the entire project lifecycle, from conceptualization to validation. Divided into seven main phases, it provides a structured approach that includes a variety of actions, which can be executed simultaneously or independently based on their classification. This flexibility accommodates the needs of different types of missions, whether complex or academic. By integrating best practices from established aerospace projects and adapting them to Latin America's unique challenges and opportunities, aerospace engineers and project managers will gain a deeper understanding and be better equipped to deliver innovative solutions.

The IPN comprehensive seven-phase framework encompasses the entire aerospace project lifecycle, enabling concurrent or independent actions to optimize project execution. By fostering collaboration, driving technological advancements, and providing innovative solutions in aerospace engineering and management, this framework enhances project outcomes. Moreover, the tailored methodology supports broader regional progress, benefiting not only Mexico but also other Latin American countries in their pursuit of aerospace technology advancement.

The methodology integrates descriptive, qualitative, quantitative, and inferential analyses to ensure robust risk management, informed decision-making, and effective configuration control. Drawing on best practices from the NASA Systems Engineering Handbook, it is tailored to the specific cultural, economic, and regulatory landscape of Latin America. Designed in Spanish, the framework fosters collaboration among multidisciplinary teams and enhances the academic and practical capabilities of students and researchers. By addressing regional challenges, the methodology promotes self-sufficiency in aerospace technology and strengthens integration into the global aerospace sector.

Dr. Pablo Alejandro Arizpe Carreón is the Academic Coordinator of The Aerospace Development Center of National Polytechnic Institute (IPN) in Mexico. He has been involved in propulsion systems of rocket and motor engines, He has more than 13 years teaching subjects related to gas turbine engines design and thermal fluids analysis. He graduated with honors from the Aeronautical Engineering program at the IPN, worked at the Mexico City airport in the helicopter maintenance area; awarded a postgraduate scholarship by the National Council of Science and Technology of Mexico (CONAHCyT) to Russia; program director of Aeronautical Engineering and creator of the aerospace master's degree at Metropolitan University of Hidalgo and has helped send more than 100 students abroad in exchange programs including NASA, Russia and France.

Dr. Jhonatan F Eulopa-Hernandez received his B.S. in aeronautical engineering from National Polytechnic Institute (IPN), México, M.S. degree in aerospace engineering from National Polytechnic Institute, México and Ph.D. degree in Autonomous Air and Underwater Navigation Systems from Center for Research and Advanced Studies (CINVESTAV), México. He has experience in routing and control of aircraft repair tasks, coordination and control of the main aircraft maintenance packages. In addition, His research interests include composite materials (polymer foams) for the manufacture of molds, positioning systems for satellite tracking using parallel robots, analysis of CubeSat nanosatellite structures, development of training satellites and guidance and control for nano launchers.

Dr. Diego Alfredo Padilla Perez is the Director of the Aerospace Development Center and holds a Ph.D. in Systems Engineering, specializing in Modeling Interdependencies in Critical Systems. He also earned a Master of Science in Systems Engineering, focusing on Reliability Analysis of the Water Supply System in a Nuclear Power Plant. He graduated with a degree in Control and Automation Engineering from the E.S.I.M.E. Zacatenco at the IPN. Dr. Padilla is an active member of the Safety, Risk Analysis, Accidents, and Systems Reliability Research Group (SARACS), a transdisciplinary team dedicated to advancing the understanding and prevention of system failures, as well as modeling mobility in megacities. Additionally, he is a member of the IPN Telecommunications Expert Network and serves as a Collegiate Professor at the Higher School of Mechanical and Electrical Engineering, Zacatenco Unit.

Dr. Carlos Couder Castañeda achieved his B.Sc. in Applied Mathematics and Computation from the Faculty of Estudios Superiores Acatlán from Universidad Nacional Autónoma de México (UNAM) in 2001. In 2004, he achieved his M.Sc. in Computer Science from Computational

Research Center from National Polytechnic Institute (IPN), and he obtained his Ph.D. in Industrial Math in 2008 from the Mexican Petroleum Institute. In 2009 he began as a postdoc in applied geophysics at the University of Pau et des Pays de l'adour, France. In 2013 he was appointed professor and researcher at Aerospace Development Center of IPN. His research interests are spread across a wide range of modeling in engineering and science, particularly in math models involved in aerospace science and technology and parallel programming.

M. C. Edgar Bibiano Valente is a researcher of the National Polytechnic Institute (IPN) in Mexico in the field of 3D scanning technology applied to the aerospace sector. With an academic background that includes a Master's degree in Manufacturing Engineering, integration of satellite structures, additive manufacturing, as well as dimensional verification and validation of CubeSat type nanosatellites. He has a solid foundation in the aeronautic field, specializing in the design of rotary wing aircraft. His professional experience as a Quality Control Engineer in a helicopter Maintenance workshop has given him a practical perspective and a deep understanding of the specific challenges and requirements of the aerospace sector. Edgar Bibiano is recognized for his passion for innovation and his ability to apply advanced technologies in the aerospace industry. During his participation he shares his experience and knowledge acquired throughout his professional career.

Safeguarding Mexico: The Strategic Role of Lance-Sciesmex in the National Civil Protection and Resilience in the Space Sector

J. Américo González-Esparza, Elsa Sanchéz-García, Maria Sergeeva, and Miroslava Barragán

Abstract

Space weather (SW) events, driven by solar activity, present significant risks to modern technological systems and critical infrastructure. In response to these challenges, Mexico has developed the National Space Weather Laboratory (LANCE) and the Mexican Space Weather Service (SCIESMEX) to monitor and mitigate the impacts of SW phenomena. These institutions, working in collaboration with the National Civil Protection System (SINAPROC), play a vital role in providing real-time monitoring and early warnings and fostering preparedness across sectors of telecommunications, navigation, energy, aviation, etc. This article discusses the foundation and tasks of LANCE and SCIESMEX, their contributions to the national civil protection framework, and efforts to strengthen Mexico's resilience to SW events, the challenges they face, and the future opportunities that lie ahead for building a more resilient Mexico. The development of a network of observational instruments allowed us to collect regional data on SW parameter variations that had been unavailable in previous decades. This, in turn, opens new research areas, making SCIESMEX and

J. A. González-Esparza (✉) · E. Sanchéz-García · M. Sergeeva
Laboratorio Nacional de Clima Espacial, Instituto de Geofísca, UNAM, Morelia, MICH, Mexico
e-mail: americo@igeofisica.unam.mx

E. Sanchéz-García
e-mail: elsasg@igeofisica.unam.mx

M. Sergeeva
e-mail: msergeeva@conahcyt.mx

M. Sergeeva
SECIHTI, Instituto de Geofísca, UNAM, Morelia, MICH, Mexico

M. Barragán
Instituto de Geografía, UNAM, Mexico City, Mexico
e-mail: miroslava@geografia.unam.mx

J. A. Ferrer Pérez and A. Froehlich (eds.), *Space Fostering Mexican Society*, Southern Space Studies, https://doi.org/10.1007/978-3-031-92617-4_5

LANCE key agents in monitoring and diagnostics of low latitude SW conditions. At the same time, the challenges securing adequate funding, expanding infrastructure, and increasing public awareness still remain. Through educational outreach, international collaborations, and the creation of tools such as the *Space Weather Recommendations Guide*, LANCE and SCIESMEX are leading the national effort to mitigate the impacts of SW.

1 Introduction

Our increasing reliance on technology in everyday life has exposed society to new challenges in today's interconnected world, particularly in the economy and industrial sectors related to space activities. Space weather (SW) refers to the disturbances in Earth's natural environment caused by solar activity, which can significantly affect space applications and terrestrial infrastructure. These disturbances can have profound implications for communication, navigation, and electric power systems, making understanding and monitoring SW phenomena essential.[1,2]

A rather well-known manifestation of solar activity is the generation of auroras, but the impact of SWx extends far beyond these natural light displays. The solar phenomena that disturb SWx conditions include solar flares, solar energetic particles (SEPs), and coronal mass ejections (CMEs) (Table 1).[3] During these events solar particles, magnetic fields, and electromagnetic radiation can interact with the Earth's atmosphere and magnetic field, which potentially can result in degrading and/or disrupting the operation of various technological systems.[4]

Usually, intense solar storms involve three phenomena: flares, SEPs, and CMEs. Nonetheless, flares and CMEs can also occur independently, and they might or might not generate SEPs.

Solar flares are sudden large explosions on the Sun with a release of high-energy electromagnetic radiation at all wavelengths. SEPs are the particles energized by flares or shock waves driven by CMEs. Solar flares and SEPs cause ionospheric disturbances that in turn can degrade telecommunications, block radio communications, affect civil aviation, military operations, and applications dependent on the precision of Global Navigation Satellite Systems (GNSS) performance.[5] CMEs are

[1] M.G. Kivelson and C.T. Russell, Introduction to Space Physics (Cambridge University Press, 1995).
[2] D.N. Baker and L. J. Lanzerotti, Resource letter SW1: Space weather, (American Journal of Physics, 84, 2016), pp. 166–180, doi: 10.1119/1.4938403.
[3] R. Schwenn, Space Weather: The Solar Perspective. Living Reviews in Solar Physics, 3(2), (2006), pp. 1–72.
[4] L.J. Lanzerotti, Space Weather Effects on Technologies, Space Weather, 1(1), (2001), pp. 1–5.
[5] M.A. Hapgood, Understanding Space Weather: The Science of Forecasting the Sun's Influence on the Earth, Routledge (2017).

Table 1 Characteristics of solar phenomena that affect space weather

Penomenon	Characteristics
Solar Flares	Sudden release of electromagnetic energy I Energy: ~ 10^{20}–10^{25} J I Occurrence: ~ 1–5 per day at solar maximum, < 1 per week at solar minimum I Classification: Type X (most intense), M, C, B. I Sun-Earth Travel Time: 8 min (speed of light)
Solar Energetic Partilcles (SEP)	Particles (mainly protons, electrons, alpha particles) accelerated by solar flares or CMEs I Energy: ~ 10^3 eV to 10^9 eV IParticles > 40 MeV pose a risk to life in space I Occurrence: They are caused in some flares and/or CMEs events. Classification: Particles with energy > 40 MeV are a risk for life in space I Sun-Earth Travel Time: ~ 15 min—a few hours
Coronal Mass Ejections (CMEs)	Explosions in the solar atmosphere releasing plasma into space at velocities of ~ 300 to 2000 km/s I Energies: ~ 10^{24} to 10^{27} J I Occurrence: ~ 3.5 per day at maximum, ~ 0.2 per day at minimum I Classification: Fast (exceeding solar wind speed) and slow (below solar wind speed); halo type for Earth-directed. I ~ 2 to 4 days. Classification: Fast (speeds greater than ambient solar wind) and slow (speeds lower than ambient solar wind); those heading towards Earth are halo type I Sun-Earth Travel Time: ~ 2 to 4 days

massive solar plasma ejections that travel as magnetic clouds through space, sometimes intersecting the Earth's orbit. When these disturbances reach the vicinity of our planet, the magnetic fields that traveled with these clouds in some cases can temporarily interconnect with the Earth's magnetic field lines, allowing solar particles to enter the Earth's space environment, eventually resulting into a *geomagnetic storm.* Only CMEs of particular physical characteristics, such as high dynamic pressure and a particular internal magnetic field configuration, can produce a geomagnetic storm. The complex physics behind these processes remains an active area of research, with many aspects still needing to be fully understood.[6,7,8] CMEs can induce geomagnetic storms that generate Geomagnetically Induced Currents (GICs), which can damage the high-voltage power grids and transmission lines, leading to widespread power outages.[9]

The periods of high solar activity follow an approximately eleven year cycle known as the solar cycle. Near the solar maximum, solar storms tend to be more frequent and intense. Figure 1 illustrates the evolution of the last five solar cycles.

---

[6] D.N. Baker, The occurrence of operational anomalies in spacecraft and their relationship to space weather, IEEE Transactions on Plasma Science, 28(6), 2007–2016 (2000).

[7] A. Pulkkinen, Space Weather: Terrestrial Perspective. Space Weather, 5(7), (2007), pp. 1–13.

[8] C.J. Schrijver and G.L. Siscoe, Heliophysics: Space Storms and Radiation: Causes and Effects, (Cambridge University Press, 2010).

[9] D.H. Boteler, Geomagnetic Hazards to Conducting Networks, (Springer, 2019).

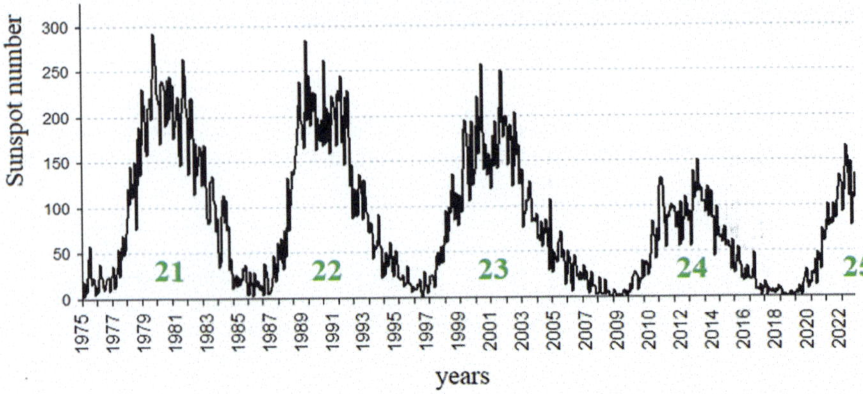

Fig. 1 Solar cycle progression in terms of the sunspot number between January 1975 and January 2024. Solar cycles 21–25 indicated by green numbers. Sunspot number is represented by it 27-day averaged value

Currently we are at the peak of solar cycle 25, during which the intense solar storms are expected. The solar activity will then gradually decrease, approaching the following solar minimum around 2030.[10]

Over the past decades, different countries documented significant effects caused by SW events. Let us mention some of the intense SW storms. The strongest solar storm in the recent history occurred on 1 September 1859 and it is called the Carrington Event in honor of the British astronomer who reported the occurrence of this specific solar flare.[11] This solar storm caused major disruptions in telegraph communications, but did not cause more catastrophic damages only because none of the modern vulnerable technologies (satellites, telecommunications, GNSS, aeronautical navigation, electric grids) existed in the nineteenth century. The SW event produced polar auroras that were observed even in Mexico.[12] Here, it is worth noting that, in general, the stronger the storm, the lower the latitudes at which the auroras could be seen. The aurora observation in Mexico means that the Carrington storm was very strong. In March 1989, a geomagnetic storm caused a nine-hour blackout in Quebec, Canada, affecting over six million people and resulting in an estimated USD 2 billion in damages.[13] In October–November 2003, a geomagnetic storm permanently damaged 15 transformers in

[10] D.H. Hathaway, The Solar Cycle. Living Reviews in Solar Physics, 12(1), 4 (2015).

[11] R.C. Carrington, Description of a singular appearance seen in the Sun on September 1, 1859, Monthly Notices of the Royal Astronomical Society, 20, (1860), pp. 13–15.

[12] J.A. González-Esparza and M.C. Cuevas-Cardona, Observations of low-latitude red aurora in Mexico during the 1859 Carrington geomagnetic storm, *Space Weather*, *16*(6), (2018), pp. 593-600.

[13] J. Allen, L. Frank, H. Sauer and P. Reiff, Effects of the March 1989 Solar Activity, Eos, Transactions American Geophysical Union, 70(46), (1989), pp. 1479–1488.

Table 2 List of some of the most intense geomagnetic storms registered in recent history

Year	Event	Sectors affected
1859	Carrington Event/solar flare-CMEs-geomagnetic storm	Telegraph systems failed; northern lights in Mexico, polar auroras observed in equatorial regions
1921	New England Central Railroad	Telegraph networks down; railroad destroyed by fire
1942	WWII Radar Outages	Radar systems failed during the war
1972	AT&T Power System Redesign	Solar flare disrupted telephone communications
1989	Quebec Blackout/ CMEs-geomagnetic storm	Six million people affected; USD 2 billion in damages
2000	Bastille Event/solar flare	Satellite short circuits; radio communications interrupted
2003	Halloween Storm/solar flares-CMEs-geomagnetic storm	Satellites affectations; communications disrupted; airlines rerouting, northern lights observed in Texas; blackouts in Sweden
2015	San Patrick Storm/solar flare	Deterioration and interruptions of communications at Cairo International Airport
2017	Solar flares-CMEs-Geomagnetic storm, a hurricane, and two earthquakes in Mexico	Combination of three different natural hazards affecting the Mexican territory
2024	Mother's Day Storm/solar flares-CMEs-geomagnetic storm	Polar auroras observed in Mexico, Portugal, Spain, Italy, Japan, South Africa, Australia, Argentina, Chile, and Uruguay; significant impact on space and power systems

South Africa's power grid.[14] More recently, on 10 May 2024, a geomagnetic storm affected multiple countries, with northern lights observed as far south as Mexico. This event affected the Mexican space sector and prompted an assessment of the national electric system's response to such disturbances,[15,16]

In our increasingly interconnected world, effective risk management associated with SW events is key to mitigating the negative impacts of SW and safeguarding critical infrastructure-free operations (Table 2).

[14] C.T. Gaunt and G. Coetzee, G. Transformer failures in regions incorrectly considered to have low GIC risk, IEEE Power and Energy Society General Meeting, (2007), pp. 1–5.

[15] J.A. Gonzalez-Esparza, E. Sanchez-Garcia, M.A. Sergeeva: et al., The Mother's Day geomagnetic storm on May 10, 2024: Aurora Observations and Low Latitude Space Weather Effects in Mexico. *ESS Open Archive.* 19 August 2024.

[16] R. Caraballo, J.A. González-Esparza, M.A. Sergeeva and C.R. Pacheco, First GIC estimates for the Mexican power grid, *Space Weather*, *18*(2), e2019SW002260 (2020).

2 Lance and Sciesmex Task

SW services monitor the conditions on the Sun, in the solar wind, in the magnetosphere, atmosphere and its parts ionosphere, and thermosphere, whose state are controlled by solar activity. These SW disturbances can significantly impact the operation and reliability of satellite-based and ground-based technological systems, posing serious risks to the everyday used technology and, as a consequence, to society.[17] Modern society's reliance on vulnerable technologies, including but not limited to radio communication networks, navigation systems, electric power grids, etc., is crucial to monitor solar activity and develop strategies to respond to potential threats.[18]

SW services of different countries work on early warnings collaborating through international organizations, for example the International Space Environment Service (ISES),[19] and the World Meteorological Organization (WMO).[20] The United Nations Office for Outer Space Affairs (UNOOSA), have actively promoted collaboration in preparedness to address SW hazards.[21] The International Civil Aviation Organization (ICAO) uses the information of global SW centers to advice airlines.[22] In addition, several nations have developed national strategies to prevent and respond to severe SW events.[23,24] Furthermore, the United Nations Office for Disaster Risk Reduction (UNDRR), under the Sendai Framework for Disaster Risk Reduction 2015–2030, recognized SW hazards in its 2020 *UNDRR/ISC Sendai Hazard Definition and Classification Review Technical Report.*[25] The International Telecommunication Union (ITU) acknowledges the importance of

---

17 P. Cannon, Extreme space weather: impacts on engineered systems and infrastructure, (Royal Academy of Engineering, Prince Philip House. 2013), ISBN 1-903496-95-0.

18 D. Baker, J.M. Bodeau, E. Cameron, et al., Severe Space Weather Events—Understanding Societal and Economic Impacts, Publisher: The National Academy Press, 641 Washington, D.C., (2008).

19 International Space Environment Service (ISES), *About ISES,* retrieved from ISES Website (all websites cited in this publication were last accessed and verified on September 2024).

20 https://community.wmo.int/en/activity-areas/wmo-space-programme-wsp/space-weather-introduction.

21 I.R. Mann, S. Di Pippo, H.J. Opgenoorth, M. Kuznetsova and D.J. Kendall, International collaboration within the United Nations Committee on the Peaceful Uses of Outer Space: Framework for international space weather services (2018– 2030), Space Weather, 16(5), (2018) pp. 428–433.

22 Doc 10100, Manual on Space Weather information in support of International Air Navigation, 1st edn, ICAO Publications, p. 586, (2019), ISBN: 978-92-9258-784-0.

23 UK Cabinet office, Space Weather Preparedness Strategy, (2015), https://assets.publishing.service.gov.uk/media/5a81521d40f0b62305b8e513/BIS-15-457-space-weather-preparedness-strategy.pdf.

24 US National Science and Technology Council (2015), Space Weather Action Plan, https://www.sworm.gov/publications/2015/swap_final__20151028.pdf.

25 Hazard definition and classification review: Technical report, United Nations Office for Disaster Risk Reduction International Science Council (ISC), (2020), https://www.undrr.org/publication/hazard-definition-and-classification-reviewtechnical-report.

utilizing the radio spectrum for SW applications.[26] Regarding research activities, there is the UN-supported International Space Weather Initiative (ISWI)[27] and the Panel on Space Weather of the Committee on Space Research (COSPAR).[28]

In 2014, in Mexico, the government amended the *General Law of Civil Protection (LGPC)* to include the SW hazards among the disruptive phenomena that the National Civil Protection System (SINAPROC) must address. These amendments emphasized the need to enhance the country's ability to anticipate and manage natural hazards, including solar and geomagnetic storms. This includes the development of comprehensive risk management strategies, SW monitoring and diagnosis, the possibility of early warnings for SW events, scientific research, and public policies, contributing directly to national security and resilience.[29]

As a result of the above, the *Mexican Space Weather Service* (SCIESMEX) was created in 2014 under the Institute of Geophysics (IGEF) of the UNAM.[30] Since 2015, SCIESMEX established a monitoring and early warning service for SW events[31] and collaborates with SINAPROC and the Mexican Space Agency (AEM). This collaboration led to the formation of the working group, *Space Weather in Mexico*, tasked with fulfilling the mandates of the LGPC. SCIESMEX is the official national source for SW information in Mexico. Figure 2 shows its relation to other entities.

The National Space Weather Laboratory (LANCE) was established in 2016 to enhance research and monitoring capabilities. One of its main tasks is the development of networks of scientific instruments to measure the set of SW parameters on a regional scale in the real time.[32] LANCE integrates the operations of SCIESMEX and develops research on solar physics, interplanetary medium, energetic particles

---

[26] Report ITU-R RS.2456-1 (09/2023), Space weather sensor systems using radio spectrum, (2023), https://www.itu.int/dms_pub/itu-r/opb/rep/R-REP-RS.2456-1-2023-PDF-E.pdf.

[27] The International Space Weather Initiative (ISWI) of the United Nations Committee on the Peaceful Uses of Outer Space was launched in 2009 and formally concluded as an agenda item of the Committee in 2012. https://www.unoosa.org/oosa/en/ourwork/psa/bssi/iswi.html.

[28] Panel on Space Weather (PSW) of the Committee on Space Research (COSPAR) https://cosparhq.cnes.fr/scientific-structure/panels/panel-on-space-weather-psw/.

[29] DOF, DECRETO por el que se reforman los articulos 2 y 82; y se adicionan la fraccion XXI del artículo 2, recorriendo el orden de las fracciones subsecuentes, y un segundo y tercer párrafos al artículo 20 de la Ley General de Protección Civil. Diario Oficial de la Federación, 6 de junio de, 1994 (2014).

[30] C.M. Denardini, S. Dasso and J.A. Gonzalez-Esparza, Review on space weather in Latin America, 3. Development of space weather forecasting centers, *Advances in Space Research*, *58*(10), 1960–1967 (2016).

[31] CENAPRED, la Agencia Espacial Mexicana y el Instituto de Geofísica de la UNAM concretan importantes logros sobre clima espacial en México: https://www.gob.mx/cenapred/articulos/cenapred-la-agencia-espacial-mexicanay-el-instituto-de-geofisica-de-la-unam-concretan-importantes-logros-de-clima-espacial-en-mexico#:~:text=2014%2D10%20Apoyo%20de%20CONACYT%20para%20conformar%20un,Creaci%C3%B3n%20del%20Servicio%20de%20Clima%20Espacial%20(SCIESMEX.

[32] J. A. González-Esparza, Space physics career in a developing country: Opportunities and challenges, *Frontiers in Astronomy and Space Sciences*, *9*, 952000, (2022).

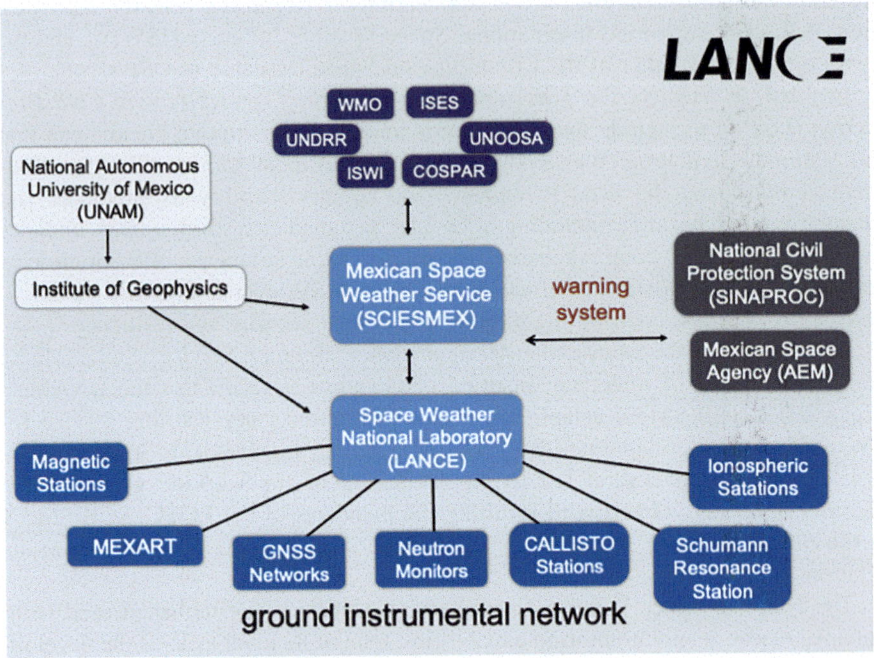

Fig. 2 LANCE-SCIESMEX organization structure in Mexico

precipitation, ionosphere, magnetosphere, and solar-terrestrial physics in general. It also studies the issues of the SW impacts on the electrical grids. Figure 2 illustrates the organizational structure of SCIESMEX-LANCE, with SCIESMEX operating under the UNAM Institute of Geophysics. The LANCE networks of observational infrastructure and instruments provide the measurements of the main regional SW parameters (Table 3).

Thanks to data provided by these instrumental networks, a thorough analysis and a real-time diagnostic of the environment over Mexico was possible during the latest SW events in September 2017 and May 2024. Figure 3 illustrates the location of different LANCE scientific instruments in the country, particularly in the states of Michoacán, Nuevo León, Mexico City, Puebla, State of Mexico, Baja California, Chiapas, Sinaloa, Quintana Roo, Veracruz, Querétaro, and soon in Sinaloa and Yucatán.

3 Space Weather Nowcasting

The SW scientific infrastructure in Mexico developed by LANCE during the last years provides the opportunity to follow the development of SW disturbances over the national territory and understand their effects. In September 2017, Mexico faced a hurricane, solar and geomagnetic storms, and two earthquakes during

Table 3 LANCE observational infrastructure of Mexico

Instrument	Function
Radio Telescope MEXART (one in operation)	Tracks solar storms in the solar wind using interplanetary scintillations measurements
CALLISTO Station Network (five instruments)	Registers solar radio bursts
Geomagnetic Station Network (REGMEX) (five instruments, three in operation)	Measures parameters of the regional geomagnetic field variations
National network of ionospheric stations (five instruments, three in operation)	Measures parameters of the regional ionosphere
Neutron Monitor Observatories (two in operation)	Measures regional solar energetic particle fluxes
All sky imagers (two instruments, one in operation)	Optical observations of ionospheric structures
Geomagnetically Induced Current (GIC) Meter Network at CFE Substations (six instruments, four in operation)	Registers GICs in the national electrical system

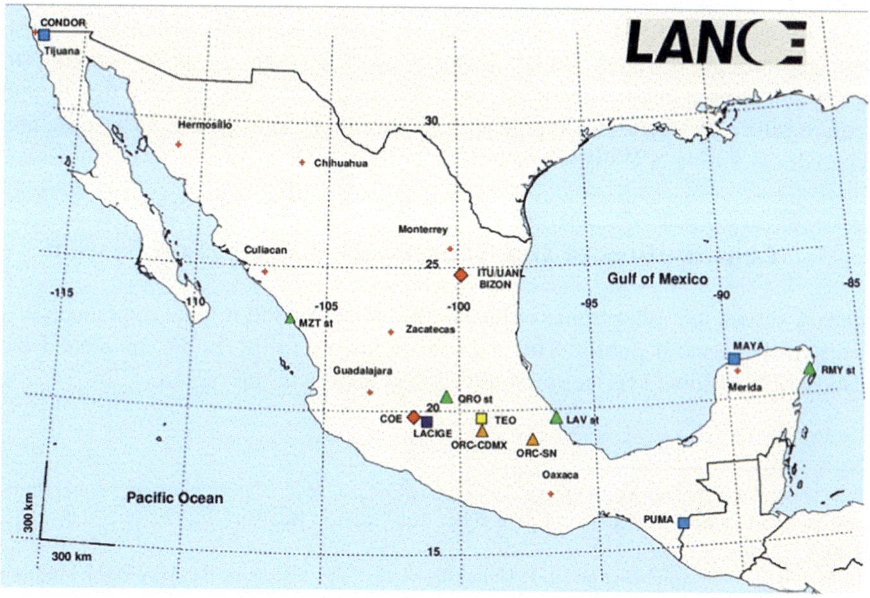

Fig. 3 LANCE's Observational infrastructure site locations in Mexico. Agenda: green triangles stand for GIC detectors in the national electric system; orange triangles neutron monitors; yellow squares magnetic stations; blue squares ionosonde stations; red diamonds indicate the Coeneo (COE), Michoacán (MIC), and Iturbide (ITU), Nuevo León observatories, which host multiple instruments including CALLISTO, magnetometers, Schumann resonances, and ionospheric stations

a short interval setting a precedent for the country's disaster management history.[33] During this period, SCIESMEX issued its first SW warning to SINAPROC. Another notable event was the geomagnetic storm of 10–12 May 2024, the most intense storm since the Halloween SWx events of October–November 2003.[34] LANCE recorded this event throughout Mexico, registrating solar chromosphere images, data on solar radio bursts, solar wind disturbances, regional variations in cosmic ray fluxes, regional geomagnetic indices, and ionospheric parameter variations. Also registered was the GIC response of the national electric system.[35]

Based on these measurements, SCIESMEX provided the necessary required scientific information important for the industry and alerts to national civil protection authorities in quasi-real time before, during and after the disturbances. This information may be used by the decision-makers in each sector during significant SW events.

In 2017, LANCE established the Institutional Repository of Space Weather (RICE) to manage the data collected from its networks. RICE is aligned with international open science principles.

In 2019, LANCE implemented a quality management system and obtained ISO 9001 certification for its *Space Weather Bulletin* preparation process which is the official mechanism to warn SINAPROC.[36]

Additionally, SCIESMEX operates an automated warning system in Spanish language through different social media platforms (such as its X account) to inform the general public about ongoing SW events.[37] This ensures timely and accurate information reaches the general public, increasing public awareness and preparedness during SW disturbances.

4 Contribution to the National Civil Protection System

Figure 4 shows the infographic published by the SINAPROC that explains SW's effects to the general public. The 2014 amendments to the LGPC mandated the creation of a national service to monitor SW's effects in the region.

---

[33] J.A. Gonzalez-Esparza, M. Sergeeva, P. Corona-Romero, et al., Space weather events, hurricanes, and earthquakes in Mexico in September 2017, *Space Weather*, *16*(12), pp. 2038–2051, (2018).

[34] L. Spogli, T. Alberti, P. Bagiacchi, P. Cafarella, et al., "The effects of the May 2024 Mother's Day superstorm over the Mediterranean sector: from data to public communication", *Annals of Geophysics*, 67(2), p. PA218, https://doi.org/10.4401/ag-9117, (2024).

[35] Gonzalez-Esparza, J. A., Sanchez-Garcia, E., M. Sergeeva, et al. (2024). The mother's day geomagnetic storm on 10 May 2024: Aurora observations and low latitude space weather effects in Mexico. *Space Weather*, 22, e2024SW004111. https://doi.org/10.1029/2024SW004111.

[36] Boletín UNAM-DGCS-352, 20 de Mayo 2019, Certifican al Servicio de Clima Espacial México, de la UNAM, https://www.dgcs.unam.mx/boletin/bdboletin/2019_352.html (2019).

[37] V. De la Luz, J.A. González-Esparza, M. Sergeeva et al., The Mexican Early Warning System for Space Weather, *Proceedings of the International Astronomical Union*, *13*(S335), pp. 236–239 (2017).

Fig. 4 National Civil Protection System SINAPROC Infographic on SW illustrating its key elements and impacts (in Spanish language)

The creation of the working group "Space Weather in Mexico" was already mentioned, which is also one of the examples of the collaborative approach to SW tasks. One of the outcomes of this effort was the publication of the "*Space Weather Recommendations Guide in Mexico*"[38] in June 2024 (Fig. 5). This official Guide contains the proposals on how to approach each of the tasks to solve and provides recommendations on which national agents should be involved in the process. The document proposes three strategic objectives, relying on collaboration between government entities and academic institutions: (1) development of scientific studies and early warning systems, (2) operation of monitoring and early warning systems and promotion of a culture of prevention, and (3) promotion of a culture of prevention.

[38] Space Weather Recommendations Guide in Mexico (2024), https://www.cenapred.unam.mx/es/Publicaciones/archivos/GUIACLIMAESPACIALRECOMENDACIONES.pdf.

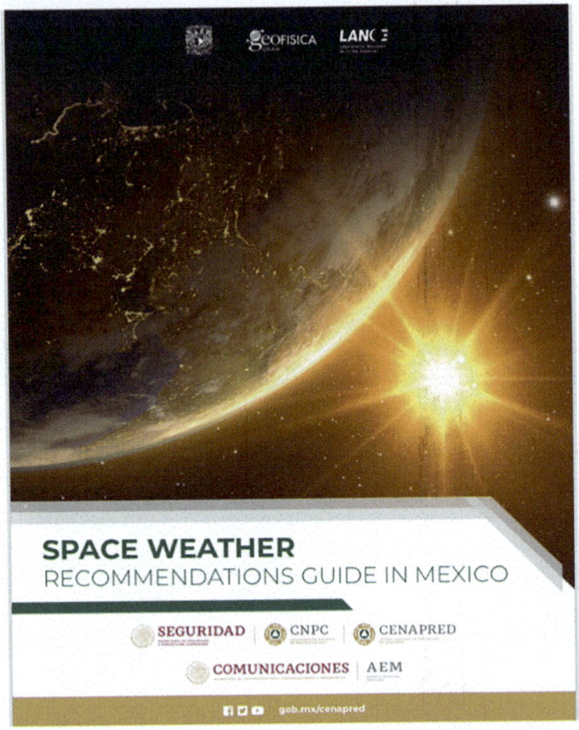

Fig. 5 Cover of the "Space Weather Recommendations Guide in Mexico", published in June 2024. This guide outlines strategies for mitigating the impacts of SW in the region

Feedback from users of SW information and decision-makers is an important issue for the SW services in developing user-targeted products and providing the most effective recommendations for different technological sectors and industry-targeted alerts.[39] The information on the degradation of performance and/or outages of specific applications and systems is crucial for understanding the potential negative impacts of SW and for improving future forecasting accuracy. However, this information is rarely disclosed by industry sectors or authorities due to confidentiality concerns. It is presumed that commercial interests prevent the reporting of such failures, even when they are caused by natural phenomena. The data sharing and collaborative efforts between the operational SW services and industry are the key issues in the SW negative impact problem-solving. In this regard, SCIESMEX promotes collaborations with entities in different sectors of the national economy, for example, with the Federal Commission of Electricity in Mexico, the Mexican Federal Institute of Telecommunications (IFT), etc.

[39] M.A. Sergeeva, J.A. Gonzalez-Esparza, V. Gatica-Acevedo et al., On Some Challenges for National and Global Space Weather Services, Remote Sensing, 15, 2023, https://doi.org/10.3390/rs15194839 (2023).

Future Work
The continued collaboration between LANCE, SCIESMEX, AEM, and SINAPROC is crucial for mitigating the impacts of space weather (SW) in Mexico. Future efforts should focus on expanding instrumental networks, enhancing data-sharing protocols, exploring advanced predictive models, fostering national and international collaborations, and increasing outreach activities on this critical topic. Additionally, establishing partnerships with industry sectors through confidential agreements will be essential to access failure data, which can further improve space weather resilience strategies.

5 Educational Outreach and Public Awareness

LANCE takes an active part in raising public awareness on SW topics through its outreach initiatives and generating content on social media. For instance, during the 10–12 May 2024, geomagnetic storm, UNAM's social communication office spread the official information on what was happening based on the reports by LANCE, which significantly impacted newspapers, radio, and electronic media. Having official sources is very important for antispoofing, as misinformation and fake news on social networks are capable of confusing and even causing panic among the public.

The *working group Space Weather in Mexico* organizes special workshops and training sessions for personnel from different sectors. These initiatives aim to prepare critical sectors for the potential impacts of severe SW events on various educational levels.[40]

The LANCE team's active involvement in undergraduate and graduate space sciences programs is of significant importance. Their participation ensures the continuous growth of expertise in SW research and monitoring in Mexico.[41]

[40] The National Center for Disaster Prevention (CENAPRED), through the National School of Civil Protection, convenes the distance course "Space Weather Risks due to Astronomical Phenomena" https://www.gob.mx/cenapred/documentos/convocatoria-2024-riesgos-de-clima-espacial-por-fenomenos-astronomicos.

[41] Examples of undergraduate and graduate space sciences programs related to SW studies can be found at: https://www.enesmorelia.unam.mx/licenciaturas/geociencias/ and https://www.pctierra.unam.mx/.

6 Challenges and Opportunities

SCIESMEX, LANCE and the SINAPROC face several challenges in advancing the comprehensive risk management for SW awareness and resilience in Mexico. However, with the larger involvement of authorities and the implementation of the proposed strategies, one can significantly mitigate the impacts of SW, offering hope for a more resilient governance, authorities, industry, and population in the future.

The instrumental infrastructure developed by LANCE presents new research opportunities. With regional data available that was previously absent, Mexico can contribute to global SW monitoring efforts and further understanding and preparedness for SW events.

7 Conclusion

SW poses a growing threat to the technological infrastructure that modern society depends on, and Mexico is no exception. The establishment of SCIESMEX and LANCE, in collaboration with SINAPROC, represents significant progress in mitigating the potential impacts of severe SW events. These Mexican institutions play a critical role in strengthening the country's preparedness and resilience through real-time monitoring, early warnings, and collaborative efforts with key sectors. Recent advancements in observational networks have provided crucial data for monitoring and warning systems, as well as historical records that aid in analyzing regional SW effects and the vulnerabilities of various technological sectors. Additionally, SW research creates opportunities for fostering international cooperation and advancing educational foresight.

Nevertheless, addressing key challenges is essential, particularly securing sufficient funding, strengthening national and international collaborations, and promoting further academic research. Success in these areas will require the coordinated efforts of authorities, industry, and educational institutions.

LANCE's advancements in data collection and analysis open new opportunities for advanced research and the continued development of SW monitoring systems. The progress achieved thus far represents a vital step toward safeguarding Mexico's future and enhancing societal resilience against the risks posed by space weather.

Dr. J. Americo Gonzalez-Esparza is a professor at the Institute of Geophysics, National Autonomous University of Mexico (UNAM). He holds a Bachelor of Science in Physics from UNAM and earned his Ph.D. in Space Physics from Imperial College London, United Kingdom. Dr. González-Esparza is a key member of the Mexican Space Weather Service (SCIESMEX) and oversees the National Space Weather Laboratory (LANCE). His research focuses on space weather, coronal mass ejections, geomagnetic storms, and solar wind disturbances.

Dr. Elsa Sanchez-Garcia is a Postdoctoral fellow at the Institute of Geophysics, National Autonomous University of Mexico (UNAM). She has a degree in Physics and Mathematics Sciences from the UMSNH, and she has a master's degree and doctorate in Earth Sciences from the UNAM. Collaborator at the National Space Weather Laboratory (LANCE). Her research topics are stream interactions regions, coronal holes, geomagnetic storm and space weather.

Maria Sergeeva is a SECIHTI Research Fellow and a researcher responsible for the ionosphere studies in the Mexican Space Weather Service (SCiESMEX). She received her PhD degree in the Saint-Petersburg State University of Aerospace Instrumentation. Her current research focus is on the ionospheric weather and disturbances and scientific ground-based instrumentation in Mexico.

Dr. Miroslava Barragan works at the Institute of Geography of the National Autonomous University of Mexico (UNAM). She has a degree in Economics from UNAM, a master's degree in Government and Public Affairs from Latin American Faculty in Social Sciences (FLACSO), Mexico and a doctorate in Economics from UNAM. She has collaborated with the *Mexico Space Weather Group*. Her research topics are public policies applied to Earth Sciences.

Satellite Development in Mexico with International Support and Its New National Initiatives

José Alberto Ramírez-Aguilar, Dulce Carolina Sánchez-Hernández, Raúl Infante-Galindo, and Jorge Alfredo Ferrer-Pérez

Abstract

Satellite technology has revolutionized communications worldwide. Most developed countries have long used the transmission of information via satellite. Mexico is no exception, as in 1968, with its integration into the INTELSAT consortium, the race to manage this technology began. However, since then, investments have been made, such as the Morelos I and II, Solidaridad 1 and 2, and SATMEX and MEXSAT satellites, solely for their use without influencing their design, manufacture, or control in space. Due to the above, in the 1990s, an initiative was presented by the National Autonomous University of Mexico through the PUIDE program, in which the efforts of Mexican students were added to the creation of the first satellite developed in Mexican territory UNAMSAT. From that moment, the opportunity and concern arose to continue working on other projects (KuauhtliSat, K' OTO, AzTechSat-1, etc.) to achieve, as a country, our own satellite technology and protection of national sovereignty.

J. A. Ramírez-Aguilar (✉) · J. A. Ferrer-Pérez
Advanced Technology Unit, School of Engineering, UNAM, Juriquilla Querétaro, Mexico
e-mail: albert09@unam.mx

J. A. Ferrer-Pérez
e-mail: ferrerp@unam.mx

D. C. Sánchez-Hernández · R. Infante-Galindo
Autonomous University of Querétaro, UAQ, Juriquilla Querétaro, Mexico
e-mail: dulce.car.san@uaq.mx

R. Infante-Galindo
e-mail: raul.infante@uaq.mx

J. A. Ferrer Pérez and A. Froehlich (eds.), *Space Fostering Mexican Society*, Southern Space Studies, https://doi.org/10.1007/978-3-031-92617-4_6

Fig. 1 Tulancingo Hidalgo Earth Station.Ultranoticias, 19 July 2024

1 Introduction

To provide satellite telecommunications services to several countries simultaneously, on August 20, 1964, the international consortium called INTELSAT (International Telecommunications Satellite Consortium) was created and signed by eleven nations, which joined together to establish a global commercial satellite communications system that would be available to all countries on a non-discriminatory basis. In 1973, the initial agreement was replaced by a permanent one when it was adopted by 54 of the meanwhile 83 member nations of INTELSAT. These nations decided that numerous political, financial, and operational benefits could be derived from pooling their resources and working together to achieve mutual space communications goals.

By 1968, unfortunately Mexico did not have special terrestrial satellite infrastructure to broadcast major world events such as the Olympic Games. The need to have a satellite system for the global transmission of the games of the modern era led the Mexican government to affiliate with the Intelsat satellite system, and the first Earth station in the country was built in the state of Hidalgo, the Tulancingo I station (Fig. 1).[1] The station rents capacity on an ATS-3 satellite owned by NASA (Fig. 2).[2]

Currently, communications through satellites are of great importance worldwide. Over the years, Mexico has begun the development of this technology due to the efforts combined economically, initially with the satellite Morelos I, obtaining the opening to the transmission of satellite communications, after which several geostationary satellites serve the country. However, it is worth mentioning that the

[1] Ultranoticias, "Proponen crear radiotelescopios con antenas de estación terrena de Tulancingo", 23 january 2025, http://ultranoticias.com.mx/proponen-crear-radiotelescopios-conantenas-de-estacion-terrena-de-tulancingo/

[2] Eospso, "Applications Technology Satellite (ATS)", 25 january 2025, http://eospso.nasa.gov/missions/applications-technology-satellite.

Fig. 2 ATS-3 satellite, NASA. Eospso, 22 October 2012

technology behind this equipment is entirely foreign. Even its operation and control are assisted, which is why, in recent years, Mexican researchers have acquired the facilities, training, and human resources necessary to make it possible for the country to generate its indigenous technology, giving rise to progress in the space field. In the future, the aim is to have satellites made and controlled by Mexican hands in orbit.[3]

2 First Generation of Mexican Satellites

Morelos Satellite System

In 1979, the Mexican government began the procedures to obtain a geostationary orbital position before the International Telecommunications Union (ITU) based in Geneva, Switzerland, for which coordination actions were carried out with other countries by the Regulations of Radiocommunications of the ITU. As a result of this coordination process in 1982, a trilateral agreement was reached between Mexico, Canada, and the United States on the orbital positions 113.5° W and 116.5° W in North America, opening the possibility of occupying them with hybrid satellites, referring to satellites that operated in the C-band (6/4 GHz) and the Ku-band (14/12 GHz). This fact allowed these positions to be registered in the International Frequency Registry of the ITU, where our country obtained powers to occupy them and international recognition and protection for their exploitation.

In this way, in November 1982, the Mexican government, through the Ministry of Communications and Transportation, commissioned its first domestic communications satellite system from the US company Hughes Communications

[3] Alvarez, C. L. (2023), Derecho satelital y del espacio exterior.

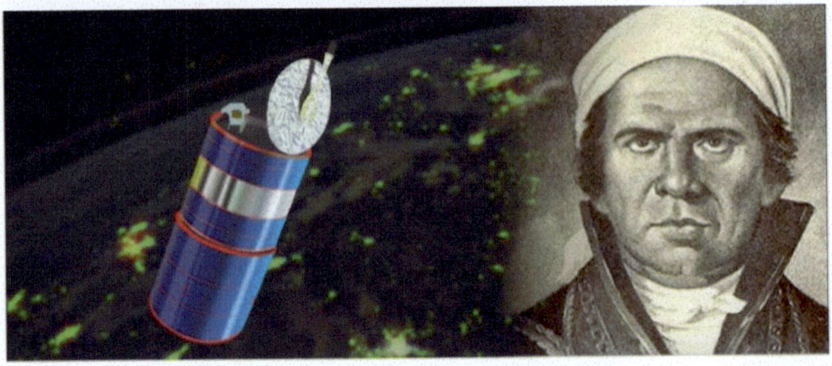

Fig. 3 Mexican Morelos Satellite System (SMS), Secretaría de Comunicaciones y Transportes (SCT)

International, a subsidiary of Hughes Aircraft Company, known today as Satellite Development Center.

The Mexican government conceived the Morelos Satellite System (SMS) to preserve sovereignty and independence in telecommunications and bring the urban and rural areas distributed throughout the vast national territory closer through its signals. In this way, the entire country would be communicated through these devices orbiting the Earth at a height of approximately 36,000 km above the plane of the equator.

On 17 June 1985, the first Mexican communications satellite, Morelos I, named after the national independence hero Don José María Morelos y Pavón, was launched into space from Cape Canaveral, Florida. It was part of NASA's space shuttle Discovery payload during mission STS-51G. Originally planned to be called "Ilhuicahua," which means "Lord owner of the sky" or "Lord of the skies," its name was changed to Morelos I in 1983 (Fig. 3).[4]

In 1985, Mexico added its name to the list of countries with their own satellites thanks to the Morelos I (Fig.4).[5] So, with the launch of this artificial satellite, Mexico entered the space age to cover the telecommunications services that the country required and that had been covered until then by the Federal Microwave Network, which was beginning to show saturation. In this way, satellite technology would start to impact the daily lives of millions of Mexicans in the national territory. After five months of maneuvers to reach its final position within the geostationary orbit, Morelos I occupied the orbital position 113.5° W on 17 December 1985. Five months later, on 26 November 1985, aboard the STS-61B mission of

[4] SCT, "Hace 35 años México tenía un satélite por primera vez en la historia", 25 january 2025, http://www.gob.mx/sct/articulos/hace-35-anos-mexico-tenia-un-satelite-por-primera-vez-en-la-historia#.

[5] SCT, "Morelos I, primer satélite mexicano", 25 january 2025, https://www.comunicacionfi.unam.mx/mostrar_nota.php?id_noticia=1473.

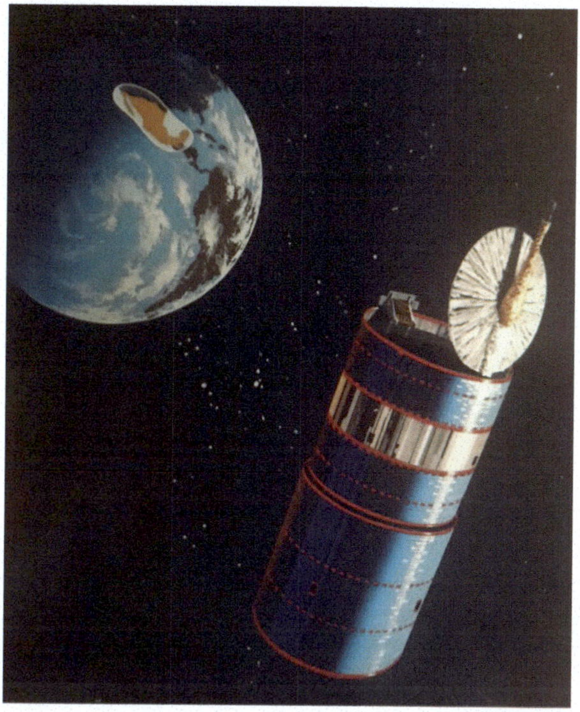

Fig. 4 1985 marks a success for Mexico's technological space development, Morelos I

the NASA space shuttle Atlantis from Cape Canaveral, the second satellite system satellite, Morelos II, was launched (Fig. 5).[6] It originally entered a storage orbit for four years to enter operations from the orbital position 116.8° W on 1 September 1989. In August 1998, Morelos II began to operate in an inclined orbit and was controlled from Mexico by Mexican engineers.

In 1989, Telecommunications of Mexico (Telecomm) was created to direct and manage the (SMS). This decentralized organization oversaw providing signal transmission services by national satellites. The country joined a select group of nations that, at that time, had their own satellite communications system.

The (SMS) provided advanced telecommunications to the entire territory, even in remote areas where no operator could enter. These services comprised educational television, commercial programs through the national television network, telephone and fax services, and data transmission through various public and private networks.

[6] Picryl, "61B-39-029-STS-61B-MORELOS satellite deploys from payload bay of Atlantis", 25 january 2025, https://picryl.com/media/61b-39-029-sts-61b-morelos-satellite-deploys-from-payload-bay-of-atlantis-ca469d.

Fig. 5 Morelos II satellite deploying from the payload bay of the orbiter Atlantis during STS-61B

México was the first client to use the Hughes 376 (Fig. 6)[7] model as a hybrid satellite operated simultaneously in two frequency bands (C and Ku). It was also the first Latin American country in which Hughes Space and Communications Company was the main contractor for a satellite project. The contract also called for the construction of the two satellites and Hughes to build and install a tracking, telemetry, and command station to operate the (SMS). The station was installed in the Iztapalapa district, formerly known as the Federal District, which was inaugurated by the then President Miguel de la Madrid Hurtado just a few days before the launch of the first satellite, that is, on 3 June 1985. It was named the Walter C. Buchanan Space Center in honor of the distinguished engineer who promoted space activities during his tenure as head of the Ministry of Communications and Transportation during the six-year term of President Adolfo López Mateos. According to the federal government, the satellite system cost 92 million USD and has an estimated useful life of nine years.

Although the Morelos satellite bus design was the standard Hughes 376, the antenna system was tailored to meet Mexico's communications needs. The four Ku-band transponders used planar arrays for receive only and had a bandwidth of 108 MHz with a minimum adequate isotropic radiated power (EIRP) of 44 dBW throughout Mexico.

The C-band transmit and receive beams and the Ku-band transmit beams are generated by a 1.8 m wide shared aperture grating antenna with two polarization-selective surfaces. The front surface was sensitive to horizontally polarized beams, and the rear surface to vertically polarized beams. Separate microwave feed networks were used for the two polarizations. The twelve vertically polarized

[7] La Jornada, "Conmemoran lanzamiento del primer satélite mexicano", 25 january 2025, http://www.jornada.com.mx/noticia/2020/06/17/economia/necesidades-de-mexico-201csiguen-siendo-gigantescas201d-espriu-9491.

Fig. 6 Morelos I satellite (Hughes-376) at the Hughes Communications International plant

narrowband transponders had a bandwidth of 36 MHz, and the six horizontally polarized wideband transponders had a bandwidth of 72 MHz. The C-band EIRP was 36 dBW throughout Mexico. The C-band narrowband transponders used 7-watt traveling wave tube amplifiers (TWTAs) with 14 by twelve redundancy. The C-band wideband transponders used 10.5-watt TWTAs with eight by six redundancy. The Ku-band transponders used 20-watt TWTAs with six by four redundancy.

Moreover, I satellite at the Hughes Communications International plant
Each satellite used a McDonnell Douglas Payload Assist Module (PAM) to boost it into an elliptical transfer orbit after release from the shuttle. Sitting in its unique cradle, the satellite used only one-seventh of the shuttle's payload bay capacity. The shuttle carried the satellite into Low-Earth Orbit (LEO), opened the cradle's protective sun shield, and rotated a turntable at its base at 55 rpm to provide gyroscopic stability. Four springs pushed the satellite into space; 45 minutes later, an onboard sequencer fired the PAM. A Thiokol Star 30B apogee motor fired on command, placing the satellite into a circular synchronous orbit. As the satellite neared its operating position, the reflector antenna and electronics platform were swung relative to Earth and achieved near-pointing accuracy using a beacon tracking system. The remainder of the bus and solar arrays continued to rotate to maintain stability. The satellite was drifted into final orbit and placed into operational position using onboard thrusters. The Morelos I and Morelos II satellites were placed at a nominal position of 113.5° W longitude and 116.8° W longitude, respectively.[8]

[8] Ramirez-Aguilar, J. A., Sanchez-Hernandez, D. C., Ferrer-Perez, J., Chávez-Moreno, R., & Romo-Fuentes, C., (2021), Satellites for the Benefit of Mexican People: Past, Present and Future. In Space Fostering Latin American Societies: Developing the Latin American Continent Through Space, Part 2 (pp. 19-35). Cham: Springer International Publishing.

Table 1 Technical characteristics of the Morelos satellites

Technical characteristics of the Morelos I y Morelos II satellites		
Name	Satelite Morelos I (Satmex I from the privatization of 1997)	Satelite Morelos II (Satmex II after privatization in 1997)
Construction Company	Hughes Space and Communications Company	Hughes Space and Communications Company
Model	Hughes HS-376	Hughes HS-376
Size	6.58m long and 2.2m in diameter	6.62m long and 2.2m in diameter
Weight	512 Kg	645 Kg
Orbit	Geostationary	Geostationary
Orbital Position	113.5° W	113.5° West, 114.5 W (migrated from March 2006)
Launch vehicle	Space Shuttle Discovery	Space Shuttle Discovery
Release date	17 June de 1985	27 de November de 1985
Launching place	Cape Canaveral, Florida USA	Cape Canaveral, Florida USA
End of useful life	1993	2004

The Morelos I satellite's helpful life was only nine years as planned, and it was deorbited on 5 March 1994. The Morelos II satellite had a nine-year design life until 1994; however, thanks to a strategy to minimize corrections to its orbit, its useful life was extended until 2004.

Some of the technical characteristics of the first generation of Mexican satellites are shown in Table 1.

On 14 June, 2004, with the last of its fuel, the Morelos II was removed from the geostationary orbit and sent into a waste orbit, where it received commands to shut down all its systems and thus be deactivated entirely. This ended a learning stage for the Mexican engineers in charge of the satellites at the Iztapalapa control center, which would be reflected in the administration of the second Mexican satellite system, the Solidaridad satellite system.

3 Second Generation of Mexican Satellites

Solidarity Satellite System

In the 1990s, Mexico already had experience managing the first domestic satellite system, the SMS. All the advantages of having its satellite telecommunications system have been revealed, so the trend towards the future would be to preserve the orbital positions granted to the country by the ITU.

In May 1988, México reached a new trilateral agreement with our northern neighbors, replacing the one from 1982. This new agreement determined that our country would occupy, with hybrid satellites in the C- and Ku-bands, three orbital positions different from those occupied by the Morelos I and II satellites: 113.5° W, 114.9° W, and 116.8° W.

In 1989, as the useful life of the Morelos satellites approached the end, the Ministry of Communications and Transportation (SCT), the Mexican Institute of Communications (IMC), and Telecommunications of Mexico (Telecomm) determined to continue the project of having with its system of geostationary satellites and replace them with a new generation of hybrid satellites, named Solidaridad 1 and Solidaridad 2.

This planned system would be the government's national security communications platform since it began to be used by the Secretariat of National Defense (SEDENA), the Secretariat of the Navy (SEMAR), the Attorney General's Office (PGR), and the National Research and Security Center (CISEN) to improve the efficiency of its operations against drug trafficking, crime, and organized crime and to aid the civilian population in case of disasters.

In this way, the federal government considered satellite communications a strategic area for the country, and they had to be controlled by the governing body, the Ministry of Communications and Transportation. In May 1990, this secretariat officially notified the public of our country's intention to launch the second generation of Mexican satellites into space, the Solidaridad System, whose coverage would be regional, that is, part of the American continent and not only the national territory, which would benefit the countries of Central America, the northern part of South America and some areas of the South of the United States of America. The International Public Tender for the supply of the Solidaridad system was published on 3 December of that same year, almost two months after this publication. In February 1991, the SCT officially revealed the contract award to the US company. Hughes Communicatios Internacional, Inc., with whom he would sign the letter of intent for constructing two model HS-601 triaxial stabilization satellites (Fig. 7).[9] These new satellites would form the second generation of Mexican satellites, the Solidaridad System. Its name was due to the National Development Plan of the administration of President Carlos Salinas de Gortari, whose name was "Solidarity, united to progress".

The contract for the construction of the Solidaridad system included the updating of the Satellite Control Center in Iztapalapa (Federal District) Fig. 8.[10] Also, for the new satellites it is required the construction of a second Alternate Control Center, which was located in the city of Hermosillo, Sonora. Likewise, it included the training of the Mexican engineers in charge of their operation, the implementation of a satellite simulator, the training of six engineers at the German Space Control Center, and academic training at the master's level for eight people in telecommunications at the University of California, Los Angeles (UCLA).

[9] Space.skyrocket, "Solidaridad 1,2", 26 january 2025, http://space.skyrocket.de/doc_sdat/solidaridad-1.htm.

[10] SCT, "Sistema de Satélites Morelos", México, 1985.

Fig. 7 Triaxial stabilization Solidaridad satellites

Fig. 8 Satellite Control Center in Iztapalapa (Mexico City). Secretaría de Comunicaciones y Transportes (SCT)

The participation of TELECOMM personnel and external advisors should also be mentioned in supervising the construction, integration, and testing during the integration and testing campaign before the launch of both satellites.

The new satellite system included these two satellites with the highest space technology, which would provide digital signal conduction services for voice, video, data, tele-hearing, and television services in the already known C- and Ku-band (same as the SMS).), in addition to the L-band, intended for the mobile satellite service, which allowed communication with land, air, and maritime transport, thus covering the national territory and our heritage sea (200 nautical miles).

On 19 November 1993, the first of the new system's satellites, Solidarity 1, was successfully launched aboard the European Space Agencys Ariane 44LP rocket from the spaceport in Kourou, French Guiana. Occupying the orbital position 109.2° W. Its estimated useful life span was 14 years, according to the fuel remaining after reaching its nominal position in the geostationary orbit.

On 7 October 1994, the Mexican satellite system was completed with the launch into orbit of the Solidaridad 2 satellite aboard the Ariane 44L rocket launched from the same port of Kourou. With a helpful life span similar to that of Solidaridad 1, this satellite occupied the orbital position of 113.0° W.

With the launch into orbit of the Solidaridad satellite system and the payload in Band "L" for mobile communications, Telecomm began providing mobile voice and data satellite communications services. This allowed the Security Agencies National (ISN) to expand its communications capacity, which would allow mobility and coverage in areas where land infrastructure was not accessible.

The L-band constituted a significant advance in satellite communications in our country since it had characteristics that made it very valuable for communications, taking into account our country's geographical and demographic conditions: high mobility and penetration.

(i) Adequate operation even in extreme weather conditions (rains, hurricanes, etc.).
(ii) It can be reused in terrestrial applications (cellular telephony).
(iii) Voice and data services can be provided with small terminals.
(iv) Allows communications, in some cases, without the antenna having to be pointed directly at the satellite (versatility of applications).
(v) Allows national coverage, including our heritage sea and the exclusive economic zone.
(vi) Mexico is the only country that has reserved the L-band for the exclusive use of the government; other countries have concessioned it to individuals or do not have satellites with L-band and faced coordination problems.

In 1996, the Memorandum of Understanding was signed to coordinate L-band between Mexico, Canada, the US, and the USSR, assigning a bandwidth of 4.4 MHz for the Solidaridad satellite system.

Like the Morelos I and II satellites, the Solidaridad spacecraft carried 18 active C-band transponders with a much higher power that allowed reception by small ground terminals. Since the transponders served various regions, they were powered by Hughes-built solid-state power amplifiers (SSPAs) that ranged from ten to 16 watts of nominal power.

Also, each satellite carried 16 active transponders in the Ku-band (four times the capacity of their predecessors) with traveling wave tube amplifiers (TWTA) of 42.5 watts of power. In addition, the Solidaridad satellites had an L-band channel to provide service to users traveling by land, sea, or air in rural areas. This service had four 21 -watt SSPAs connected in parallel.

With the Solidaridad system, Mexico increased the signal conduction capacity by having more excellent power amplifiers; the terrestrial infrastructure cost was reduced by requiring smaller earth stations, and the country entered into international market competition due to its multiple regional footprints. Besides this, at the end of 1994, the satellite system provided data, voice, video, and audio signal conduction services to many companies in the financial and commercial sectors, in addition to carrying signals to support education, culture, science, health, and national security. In summary, the capacity of the Solidaridad system was twice that of Morelos, and its power was estimated to be around 8 times greater.

The Solidaridad I satellite replaced its predecessor, Morelos I, while Solidaridad II replaced Morelos II in terms of the number of satellites owned by the Mexican government. Originally, Morelos I was operated at the geostationary orbital position 113.5° W. However, its replacement (Solidaridad II) was later placed half a degree to the East, so Mexico occupied the 113.0° W position. The Morelos II satellite, the second of this first generation, was placed in its nominal orbital position at 116.8° W, beginning its operations on 1 September 1989.

Even though the estimated useful life for the Solidaridad 1 satellite was until 2007, it recorded failures in its flight computer in 1999 and 2000 that put at risk the future use of operational frequencies and bandwidths for the region, so on 29 August 2000, it was exorbitant. Likewise, Solidaridad 2's helpful life would end in 2008. Still, when it was placed in an inclined orbit, it operated from 2008 to 2013, providing service to the government. It was the only satellite that could operate encrypted signals, voice, and data at the service of the Ministry of Defense: the Attorney General's Office, the Ministry of Public Security, and the Mexican Navy. Solidaridad II was deorbited in the early hours of 30 November 2013, already part of the company Satélites Mexicanos (SATMEX) assets. Some of the technical characteristics of this generation of satellites are shown in Table 2.

4 SATMEX—Third Generation of Mexican Satellites

SATMEX

By the year 1983, and due to the imminent launch into orbit of the SMS, it was established in the Political Constitution of the United Mexican States, specifically in article 28, that satellite communication would be a restricted area in which the provision of the satellite signal conduction service remained in the hands of the Mexican State. Such reform to the Constitution was published in the Official Gazette of the Federation (DOF) on 3 February 1983, with Lic. Miguel de la Madrid is the president of the republic.[11]

[11] Álvarez C. L. (UNAM Posgrado).2018. Telecomunicaciones y Radiodifusión en México.

Table 2 Technical characteristics of the Solidaridad satellites

Technical characteristics of the Solidaridad I and Solidaridad II satellites		
Name	Satelite Solidaridad I (Satmex III after privatization in 1997)	Satelite Solidaridad III (Satmex IV after privatization in 1997)
Construction Company	Hughes space and communications company	Hughes space and communications company
Model	Hughes HS-601	Hughes HS-601
Size	3 m × 7 m	3m × 7m
Weigth	2776 Kg	1670 Kg
Orbit	Geoestacionaria	Geoestacionaria
Orbital position	109.2° West	113.5° W
Launch vehicle	Ariane- 44LH 10 +	Ariane- 44LH 10 +
Release date	19 November de 1993	7 de October de 1994
Launching place	Kourou, Guyana Francesa	Kourou, Guyana Francesa
End of useful life	14 years (In 2000 it failed, so it was decommissioned before reaching its scheduled useful life in 2007)	14 years (After a severe failure it was transferred to an inclined orbit in March 2008)

For around 14 years, the exploitation of satellite communications in our country was the exclusive preserve of the Mexican Government, but that began to change in January 1995 when the Federal Executive presented a bill to reform once again the fourth paragraph of article 28 of the Political Constitution of the United Mexican States on satellite communications. This modification meant that satellite communication in our country would cease to be a "strategic area" and become a "priority," in which private capital would be allowed to participate in the ownership and operation of satellite communications in Geostationary orbit.[12]

The Mexican satellite system, thanks to the initiative of the State and the reform of Article 28 of the Political Constitution of the United Mexican States, took a definitive direction when, on 7 June 1995, the Federal Telecommunications Law was issued, which opened the private investment and the possibility of granting concessions through the public bidding procedure to Mexican companies was established, with a foreign participation limit of 49% to occupy geostationary orbital positions and satellite orbits assigned to the country for the provision of satellite services. The Mexican satellite system was administered since 1989 by the decentralized body of the federal government "Telecommunications of Mexico" (TELECOMM); at the beginning of 1997, it managed three satellites in geostationary orbit: The Morelos II satellite and the Solidaridad 1 and Solidaridad 2 satellites. By then, the new Mexican satellite that would replace Morelos II was

[12] Álvarez C. L. (UNAM Posgrado).2018. Telecomunicaciones y Radiodifusión en México.

Fig. 9 Mexican Satellite, S.A. of C.V. (SATMEX).

already under construction, ultimately named SATMEX 5. The privatization process of the Fixed Satellite Services Section of TELECOMM, which had begun in 1995 with the reform of the Constitution and subsequently the issuance of the Federal Telecommunications Law, was completed in June 1997 with the creation of the company Satélites Mexicanos, S.A. de C.V. (SATMEX). Figure 9 shows the logos used to promote the satellite.[13]

This company was formed with the participation of 75% private capital and only 25% government participation without voting rights, in which the concessions of the three orbital positions that Mexico were assigned, in addition to the Morelos II and its next replacement (SATMEX 5), Solidaridad I and II, as well as the two control centers (CDMX and Hermosillo, Sonora). Its objective was to provide television, voice, and data signal conduction services for public and private communication networks, teleports for the provision of social services such as educational and cultural television, the transmission of special events, rural telephone services, and support to the telegraph network, occupying less than 10% of the satellite capacity. Figure 10 is an image of the satellite control center in D.F., now CDMX.

The creditor consortium of the international public tender was formed by the Mexican company Telefónica Autrey, S.A. de C.V. and the US company Loral Space & Communications, Ltd. The Federal Government retained a reserve of satellite capacity equivalent to 7% of the total capacity. Through TELECOMM, it retained the teleports and the L-band mobile satellite service to provide mobile services, mainly for national security entities.

The Mexican State's participation in the SATMEX company was reduced over time. In 1999, it was reduced to 23.6%, in 2006 to 4%, and in 2011, the Federal Government was no longer a shareholder of the company.

Already established under Mexican laws, SATMEX was a company that provided the following services[14]:

(i) Fixed satellite communication services to clients who, in turn, provided fixed, cellular, and rural telephone services, internet, distance education, and control of maritime operations.

---

[13] Crunch Base, "Satmex Logo", 22 january 2025, https://www.crunchbase.com/organization/satmex.

[14] Álvarez C. (UNAM Posgrado).2018. Telecomunicaciones y Radiodifusión en México.

Fig. 10 SATMEX Satellite Control Center in Iztapalapa (Mexico City). *Source* own Research

(ii) Satellite capacity for radio and television transmissions.
(iii) Operated a business line in the USA called Alterna TV, which was created to offer programming in Spanish through the pay-TV system.

Satélites Mexicanos was the holder of three concessions to exploit orbital positions assigned to the country with their respective frequency bands and emission and reception rights, occupation of the geostationary orbital positions allocated to the government, granted by the SCT on 23 October 1997. These concessions authorized it to provide satellite capacity to concessionaires and permit holders through the C- and Ku-bands associated with the geostationary positions 109.2° W, 113.0° W with national coverage, and 116.8° W with continental coverage. C- and Ku-bands are used for fixed services, while L-bands are used for mobile services.

For the year 2007 and due to the proximity of the end of the useful life of the Solidaridad 2 satellite, the Mexican government requested SATMEX to put it into inclined orbit starting in March 2008 to prolong its useful life and continue to have the mobile services in the L-band, even if communications in the C- and Ku-bands were no longer used. Fortunately, thanks to the administration of the control center, the Solidaridad 2 satellite could continue providing national security communications. At the same time, the new Mexican Satellite System project progressed, and its first satellite was launched into space in 2012.

SATMEX 5 Satellite

In August 1996, Hughes Space and Communications International, Inc. and Boeing Satellite Systems International, Inc. signed a contract with SATMEX, formerly

Fig. 11 SATMEX 5 occupied the orbital position 116.8° W.

TELECOMM, for the construction of the Morelos 3 satellite, later renamed SATMEX 5 (Fig. 11), (as it was the fifth Mexican satellite to be launched), a Hughes HS-601HP satellite with a triaxially stabilized body.[15]

It was successfully launched on 5 December 1998 from the Kourou spaceport, French Guiana by an Ariane-42L H10-3 (Ariane 5) rocket. This satellite replaced Morelos II at an orbital position of 116.8° W.

SATMEX 5 was a Hughes HS-601HP or "high power" version of the body-stabilized spacecraft. It had more than 7000 watts of payload power, at least ten times the capacity of the Morelos II. Thanks to innovations such as double-junction gallium arsenide solar cells, radiation-cooled traveling wave tube amplifiers (TWTA), advanced battery technology, and a xenon ion propulsion system (XIPS) the HS-601HP (Fig. 12) could carry payloads twice as potent as those of the HS 601 models, with an expected useful life of 15 years.[16]

With an improved design, SATMEX 5 provided business communications, television distribution, and educational programming services, among others, with great reliability, performance, and value. In addition, within the Ku-band geographic coverage area, direct-to-home services were provided with 60 centimeters or fewer antennas. The worst-case EIRP and G/T (EIRP (Effective Isotropic Radiated Power), G/T (The G/T ratio essentially compares the receiving antenna's gain to the system noise temperature) margins (are sufficient for digital broadcasting. The satellite beams expanded coverage to the entire American continent, providing services to countries such as Argentina, Brazil, Chile, Colombia, El Salvador, Peru, Venezuela, and the United States.

SATMEX 5 was built at Hughes Space and Communications Company's Integrated Satellite Factory in El Segundo, California. As in the previous Morelos and Solidaridad programs, the work team comprised personnel from Hughes and

[15] Sateliitesdemexico, "Satélites de México y algo más", 22 january 2025, http://satelitesdemexico.blogspot.com/2008/09/satmex-5html.

[16] Gunter's Space Page, "SATMEX 6 → Eutelsat 113 West A", 22 january 2025, https://space.skyrocket.de/doc_sdat/satmex-6.htm.

Fig. 12 Illustration of the SATMEX 5 Satellite, HSCI model HS-601HP

Mexico. Our country sent more than ten engineers to work on the design and integration of the spacecraft.

Unfortunately, SATMEX 5's XIPS propulsion system failed on 27 January 2010, so the chemical propulsion mechanism had to be installed and used. However, it only lasted 2.5 years. Finally, at the end of 2013, its operational life ended after it was placed in an inclined orbit due to the thrusters' failure.

The European operator Eutelsat acquired SATMEX in 2014 and renamed it Eutelsat Américas. In May 2014, SATMEX 5 was renamed Eutelsat 115 West A.

The satellite was retired in 2016 and moved to a graveyard orbit above the geostationary belt.

SATMEX 6 Satellite

Derived from the imminent need to replace the Solidaridad 1 satellite in the geo-orbital position 109.2° W, the corresponding negotiations were initiated with the government of Canada to foresee possible interference to its satellites located in positions 107.3° W and 111.1° W, both adjacent to position 109.2° W in 25 August 2003 (Fig. 13).

Fig. 13 Geo-orbital positions for Canada and Mexico before the exchange, Regulación Satelital en México. Estudios y acciones (COFETEL), 1995

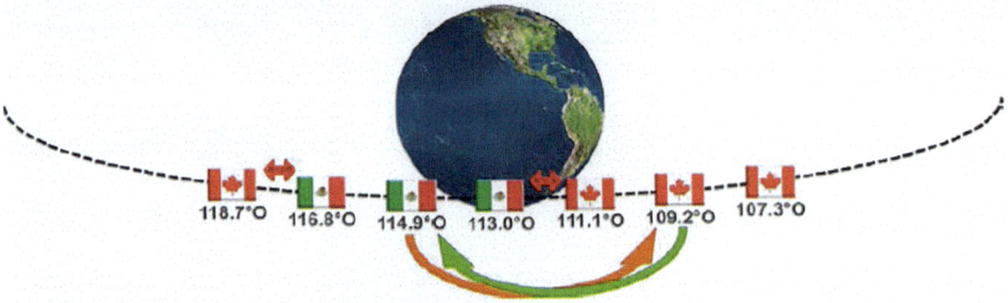

Fig. 14 Geo-orbital positions for Canada and Mexico after the exchange, Regulación Satelital en México. Estudios y acciones (COFETEL), 1995

Despite the efforts of both administrations to coordinate the new satellite that would replace Solidaridad 1, it turned out to be technically impossible, so it was decided to change orbital positions between both countries.

The exchange of the orbital positions 109.2° W belonging to Mexico and 114.9° W belonging to Canada allowed the three Mexican orbital positions to be grouped in order to minimize the potential interference between the satellites of both countries. This exchange of positions was formalized through a "Memorandum of Understanding" signed on 25 August 2003 by the governments of both countries (Figs. 13 and 14).[17]

The exchange of orbital positions allowed the relocation of the Solidaridad 2 satellite to position 114.9° W and, after the launch of the SATMEX 6 satellite in May 2006, its placement in position 113.0° W (Fig. 15).[18]

SATMEX 6 was a model SSL-1300X (Fig. 16), a satellite built by Space Systems/Loral at the SS/L facilities in Palo Alto, California. Launched on 27 May 2006 from the Kourou spaceport, French Guiana, carried into space by an Ariane 5 ECA rocket from the Arianespace company.[19]

At takeoff, the satellite weighed approximately 5500 kg. It was positioned at 113.0° W and carried 36 C-band and 24 Ku-band transponders.

Satélites Mexicanos used the state-of-the-art SATMEX 6 satellite to reinforce its telecommunications and Internet transmission capacity for Mexico, other parts of North America, and Latin America.

SATMEX was acquired by Eutelsat in 2014 and was renamed Eutelsat Américas. In May 2014, SATMEX 6 was renamed Eutelsat 113 West A.

[17] IFT, "Regulación satelital en México", 28 january 2025, https://www.ift.org.mx/sites/default/files/contenidogeneral/espectro-radioelectrico/regulacionsatelitalenmexicoestudioyacciones19-06-2013-final.pdf.

[18] Word Press, "Satelites en Latinoamerica", 23 january 2025, https://satelitesamerica.wordpress.com/2017/02/16/satmex-6-eutelsat-113a/

[19] Space.Skyroket, "Satmex 5 Eutelsat 115 west A", 27 january 2025, http:/space.skyrocket.de/doc_sdat/satmex-5htm.

Fig. 15 SATMEX 6 occupied the geo-orbital position 113.0° W

Fig. 16 Illustration of the SATMEX 6 Satellite, model SSL-1300X

Due to anomalies detected in its nominal operation, the satellite ceased operations on 31 January 2024, so it was deorbited on 3 April 2024.

With the launch of the SATMEX 5 and SATMEX 6 satellites into orbit, the coverage and expansion of the satellite market increased overwhelmingly, going from domestic coverage with Morelos I and II to Continental coverage. Figure 17 shows the evolution of the coverage of Mexican satellites.[20]

SATMEX 7 Satellite

In June 2008, Satélites Mexicanos issued an Authorization to Proceed with constructing the SATMEX 7 satellite to Space Systems/Loral (SS/L). The new satellite would provide Fixed Satellite Services (FSS) on the American continent.[21]

---

[20] IFT, "Regulación satelital en México", 28 january 2025, https://www.ift.org.mx/sites/default/files/contenidogeneral/espectro-radioelectrico/regulacionsatelitalenmexicoestudioyacciones19-06-2013-final.pdf.

[21] Romero F. 2022, Hacia la Formulación de una política espacial en México.

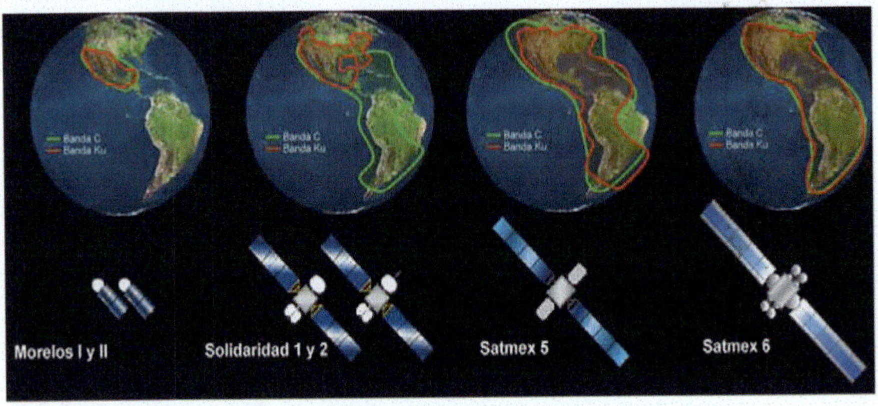

Fig. 17 Evolution of Mexican satellite coverage from geostationary orbits, Regulación Satelital en México. Estudios y acciones (COFETEL)

SATMEX 7 would be a next-generation high-power satellite in the C- and Ku-bands to increase SATMEX's fleet in the provision of services such as broadband, voice and data transmission, and video broadcasting, including high-definition television. It was originally intended to be located at 114.9° W as a replacement for the Solidaridad 2 satellite scheduled to be launched in 2011, which was about to end its useful life.

Unfortunately, the Mexican government canceled its construction after declaring a couple of tenders void, thus losing the 109.2° W orbital position.

SATMEX 8 Satellite

Space Systems/Loral (SS/L) announced in May 2010 that it had been selected to supply a new satellite to Satélites Mexicanos called SATMEX 8, which was a high-power satellite operating in the C- and Ku-bands that replaced the SATMEX 5 satellite. SATMEX 8 is to provide Fixed Satellite Services (FSS) in both North and South America.

Successfully launched on 26 March 2013, by the operator International Launch Services (ILS) aboard a Proton-M Briz rocket from the Baikonur Cosmodrome in Kazakhstan.

The new satellite increased SATMEX's current fleet in providing services such as broadband, voice and data transmission, and video broadcasting. SATMEX 8 is a satellite with 24 transponders in the C-band and 40 transponders in the Ku-band. It is designed to be located at the orbital position 114.9° W. The SATMEX 8 satellite is based on the high-power space bus SSL-1300 from SS/L (Fig.18), offering flexibility for various applications. Its expected useful life span is 15 years.[22]

[22] Space.skyrocket, "Satmex 6 Eutelsat 113 West A", 28 january 2025, http:// space.skyrocket.de/doc_sdat/satmex-6.htm.

Fig. 18 Illustration of the SATMEX 8 Satellite, model SSL-130

The sale of Mexican Satellites to Eutelsat

The constitutional reform of 11 June 2013 in telecommunications opened the door to 100% foreign investment in telecommunications and satellite communication and up to 49% in radio broadcasting.

Based on this new telecommunications reform Eutelsat Communications acquired SATMEX on 1 January 2014. Eutelsat Communications began operating under the name Eutelsat Americas starting in March 2014, and SATMEX satellites were renamed Eutelsat satellites. This company operates under the name Eutelsat Group, with headquarters in Paris, France.

At the time of the purchase by Eutelsat, SATMEX had three orbital positions occupied by satellites:

(1) SATMEX 5, renamed EUTELSAT 113 West A (113.0° W)
(2) SATMEX 6. renamed EUTELSAT 115 West A (114.9° W)
(3) SATMEX 8, renamed EUTELSAT 117 West A (116.8° W).

SATMEX satellites covered 90% of the population of America, and having an 11% share in the Latin American satellite services market.

5 Satellite Mexican System

MEXSAT

As a result of the deorbit of the Solidaridad 1 satellite and the new position of the Solidaridad 2, it was necessary to replace the satellite system ultimately since the communication it allowed would be affected and would have significant social, economic, and national security repercussions. Therefore, on 20 August 2009, the Program for National Security 2009–2012 was published in the DOF to guide the Federal Public Administration's efforts in conserving security for the benefit of the Mexican State.[23] In this document, it was foreseen that the SCT would implement

[23] Secretaría de Comunicaciones y Transportes - Sistema Satelital Mexicano, Noviembre 2015.

the new Mexican Satellite System (MEXSAT) to guarantee the availability of satellite infrastructure that would contribute to covering the needs of national security and social coverage of the state, for which the strategy of "establish a comprehensive information system for the preservation of national security" and as a line of action "ensure the availability of optimal satellite capacity and continuous service required by intelligence tasks for national security."

The rights obtained by Mexico for the aforementioned orbital positions and the use and exploitation of the associated frequency bands, among which is the L-band, whose importance for the country is paramount, are conditioned to its occupation and use by article 11.47–11.50 of the Radio Regulations, the international regulation on the subject of the ITU, and the 1996 Memorandum of Understanding, for the coordination of the L-band between Mexico-Canada-USA-Russia-United Kingdom (Inmarsat).

The useful life of the Solidaridad 2 satellite was estimated to end at the end of 2008. Given the imminence of this event, the Mexican Government began the 2007 evaluation of alternatives to guarantee the continuity of services in the short, medium, and long term. Satellites, considering the financial situation of SATMEX, the relevance of the L-band, and the preservation of orbital positions for Mexico.

As already mentioned, the federal government asked SATMEX to put the satellite in an inclined orbit starting in March 2008, thereby allowing the continuity of L-band services by extending the useful life of the Solidaridad 2 satellite, even though this limited the possibility of using the C- and Ku-bands.

The MEXSAT system, with a modern and maximum-security communications infrastructure, represents an advance in the country's satellite communications. Among the benefits of the MEXSAT System, the preservation of orbital positions and associated frequencies in favor of Mexico stands out; the provision of secure and reliable communications throughout the national territory, including the territorial sea and the exclusive economic zone, and the control and autonomy of the federal government over strategic national security satellite communications in fixed and mobile environments, in turn strengthening the position international of Mexico in satellite matters.

On 29 March 2012, during the LV General Assembly of the National Chamber of the Electronic, Telecommunications and Information Technology Industry, CANIETI, which took place in Mexico City, now known as the Secretary of Communications and Transportation, Dionisio Pérez -Jácome Friscione, announced the names of the satellites that would make up the new Mexican Satellite System MEXSAT:[24]

(1) MEXSAT 1 (Centenario)
(2) MEXSAT 2 (Morelos 3)
(3) MEXSAT 3 (Bicentenario).

[24] Telecomm Mexsat - Memoria Documental. Abril 2024.

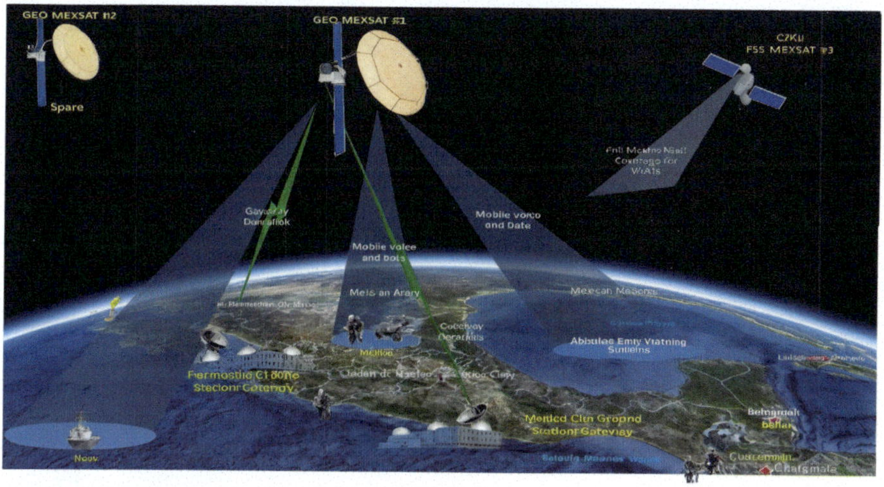

Fig. 19 Mexican Satellite System MEXSAT, TELECOMM

The new generation of Mexican satellites consisted of these three satellites: the Bicentenario satellite was for the Fixed Satellite Service, while the Centenario and Morelos 3 were intended for the Mobile Satellite Service (Fig. 19).[25]

There are two control centers for the administration of the MEXSAT Satellite System, one in Mexico City and the other in Hermosillo, Sonora. The Mexican Satellite System aims to provide robust satellite communication services that can provide services for national security, public safety, and civil protection (for example, in cases of natural disasters), and social services (for example, schools and medical brigades), among others.

MEXSAT 1 Satellite (Centenario)

Centenario was projected to be the primary satellite for the mobile service of the MEXSAT network (Fig. 20), built by the American company Boeing Satellite Systems International Inc., model Boeing 702 HP, which would be placed in orbit 113° W, in the last quarter of 2013.[26]

Together with its twin satellite, Morelos 3, both mobile satellites generate 14 kilowatts of power and include a 22 m L-band reflector. They are used to provide mobile satellite services, mainly for national security agencies.

Finally, it was launched on 5 May 2015 from the Baikonur Cosmodrome in Kazakhstan aboard a Proton M rocket. This satellite would have an estimated useful life of 15 years. However, an accident approximately 490 seconds after

---

[25] Mexicoaeroespacial, "Sistema Satelital Mexicano (MEXSAT)", 29 january 2025, http://mexicoaeroespacial.mx/2016/11/06/sistema-satelital-mexicano-mexsat/.

[26] Haciaelespacio, "Todo listo para el lanzamiento del satélite Centenario", 29 january 2025, http://haciaelespacio.aem.gob.mx/revistadigital/articul.php?interior=253.

Fig. 20 Satellite Centenario (MEXSAT 1)

Proton M took off prevented the mission from being completed. The loss of Centenario would be remedied with the subsequent launch into orbit of its twin satellite, Morelos 3.

The Federal Government contractor was Boeing Satellite Systems International Inc., which was responsible for integrating the MEXSAT network, which included two ground telemetry and control centers, associated operations systems and reference terminals for users, in addition to the three satellites. Boeing subcontracted the company Orbital Sciences Corporation, responsible for providing the ground command and control equipment, software, and associated training and operation documents, in addition to the construction of the Bicentenario satellite.

MEXSAT 2 Satellite (Morelos 3)

It was originally planned as the secondary satellite for the mobile service of the MEXSAT network, built by the American company Boeing Satellite Systems International Inc., model Boeing 702 HP, to be placed in orbit 116.8° W and launched into space in the third quarter of 2014. However, its launch was postponed.

When the twin satellite MEXSAT 1 (Centenario) was destroyed during its unsuccessful launch in 2015, Morelos 3 became its replacement.

It was finally launched on 2 October 2015 on a Lockheed Martin Commercial Launch Services Atlas-V 421 rocket from the US Air Force Base in Cape

Fig. 21 Morelos III satellite (MEXSAT 2) in the construction plant

Canaveral, Florida. After two hours and 51 minutes of flight, the Atlas-V successfully released Morelos 3 to begin its autonomous ascent to its orbital position at 36,000 kilometers altitude, which was reached in 13 days of travel.

This satellite will have an estimated useful life span of 15 years.

Figure 21 shows the Hughes Space and Communication Company plant constructing the Morelos 3 satellite, and Table 3 presents its principal technical specifications.[27]

MEXSAT 3 Satellite (Bicentenario)

The Bicentenario (Fig. 22) was the first of the MEXSAT network satellites to be put into orbit. It was manufactured by Orbital Sciences Corporation, a partner of Boeing Satellite Systems International Inc., builder of the Centenario and Morelos 3 satellites.[28] The hybrid communications satellite C- and Ku-band are based on Star-2, Orbital's flight-proven platform. Technical specifications are shown in Table 4.

[27] a21, "Morelos 3 concluye con éxito etapa de despliegue", 29 january 2025, http:// a21.com.mx/aeroespacial/2015/11/03/morelos-3-concluye-con-exito-etapa-de-despliegue.

[28] SCT, "Nuevo satélite del gobierno mexicano, listo para ser puesto en órbita", 29 january 2025, https://www.sct.gob.mx/despliega-noticias/article/nuevo-satelite-del-gobierno-mexicano-listo-para-ser-puesto-en-orbita/.

Table 3 Technical characteristics of the Centenario and Morelos 3 satellites of the MEXSAT system

Characteristics of the Centenario and Morelos 3 Satellites (Mobile Satellite Service)	
Manufacturer	Boeing
Satellite model	Boeing 702HP
Orbital position	113.0° y 116.8 W
Orbit type	Geostationary (6° inclined)
Useful life	Approximately 15 years
Antennas	22 m reflector for L-Band 2 m antenna for planned Ku Band
Coverage	Mexico, Exclusive Economic zone, Central America, South America
Total ground weight	Launch mass: 5400 kg Mass in orbit: 3200 kg
Payload	Solid State Power Amplifier for L-Bnad 100 W Traveling Wave Amplifier Tubes for Ku-Band Flexible digital channel
Power in watts	14,000
Release date	05/15/2015 (Centenary) – 10/02/2015 (Morelos 3)
Launch place	Baikonur, Kazakhstan – Cape Canaveral, Florida

Fig. 22 Bicentenario Satellite (MEXSAT 3) in the construction plant

Table 4 Technical characteristics of the Bicentenario satellite of the MEXSAT system

Characteristics of the Bicentenario Satellite (Fixed Satellite Service)	
Manufacturer	Orbital
Satellite model	Star 2
Orbital position	114.9°
Orbit type	Geostationary (Geosynchronous)
Useful life	Approximately 15 years
Propulsion	Propelente
Coverage	Mexico, Exclusive Economic Zone
Payload	12 Active Transponder Extended "Ku" Band 12 Active Transponder Extended "C" Band
Power in watts	3500 (Approximately)
Release date	19/12/2012
Launching place	Kourou, Guyana Francesa

It was launched on 19 December 2012 from the ELA-3 ramp of the European Space Agency's spaceport in Kourou, French Guiana, aboard an Ariane 5-ECA rocket. It occupies an orbital position of 114.9° W.

6 New National Initiatives in the Satellite Area

University Program of Space Research and Development

Due to the great need for Mexico to develop its satellite technology and the lack of a public policy on satellite matters, the academic sector, through the National Autonomous University of Mexico created 1992 the University Program for Space Research and Development—PUIDE (Fig. 23). There was undoubtedly Mexico's desire to integrate not only in the development of satellites but also in the development of its space technology and reduce foreign technological dependence. Likewise, the program's objective was to train qualified personnel who could face the challenges of the space technology sector in Mexico.

Gianfranco Bisiacchi Giraldi† was one of the directors of PUIDE and accepted in 1992 the coordination of the UNAMSAT microsatellite project, which was intended to orbit the Earth to detect meteoritic traces using radio Doppler reflection. The first university microsatellite was launched from Plesetzk, Russia, and began operations from space and sent radio frequency signals (telemetry) to the Earth station that was installed on the roof of the Satellite Integration Laboratory in the Instrument Center in Ciudad Universitaria, Mexico.

Bisiacchi Giraldi was a researcher who believed in the talent of young Mexicans students and engineers and took responsibility for developing space technology at UNAM. Unfortunately, Bisiacchi Giraldi died in 2006. Still, his work and support for young students, aware of the need to develop satellites in Mexico, has yielded results, as there is now a Advanced Technology Unit – UAT, school

Fig. 23 Some members of the space program—PUIDE *Source* Own Research

Fig. 24 Gianfranco Bisiacchi Giraldi†

of Engineering at UNAM where various space university projects are developed. Thank you very much Dr. Gianfranco Bisiacchi Giraldi for all your knowledge of the service of Mexico and Support for young Mexicans Students (Fig. 24).[29]

The microsatellite project continued its development, achieving the training of the first space engineers specializing in the design and construction of satellites in Mexico. With this event in 1996, UNAM became one of the first universities in the world to put an object in space (Fig. 25).[30]

[29] Scielo, "GIANFRANCO BISIACCHI GIRALDI 1947–2006", 30 january 2025, https://www.scielo.org.mx/pdf/rmaa/v43n2/v43n2a13.pdf.

[30] Dgcs, "El satélite universitario UNAMSAT-B fue puesto en órbita exitosamente", https://www.dgcs.unam.mx/boletin/bdboletin/multimedia/WAV210903/728(7).jpg.

Fig. 25 UNAMSAT-B satellite successfully enters orbit (DGCS—UNAM)

The UNAMSAT project left its foundations at the university and was projected internationally to be at the forefront of space technology. It is expected that the future in this strategic technological sector for the country will be much better for the new generations since several universities in Mexico are currently beginning to develop projects under the CubeSat standard because, with this type of structure, there is a greater probability of obtaining economic funding for its development, launch, and placement into orbit.

Several students who participated in the UNAMSAT project continue working in private industry and universities, developing space and aeronautical systems for Mexico's benefit (Fig. 26). The UNAMSAT project left evidence of the tremendous effort required for universities to establish a satellite project without the economic and logistical support of industry and the government.

Nanosatelite Tubesat Kuauhtlisat—Reaching the STARS

The nanosatellite KuauhtliSat means "Eagle" in Spanish because the UNAM shield includes an eagle, which is related to the identity of the Mexican people (Fig. 27). The nanosatellite has served as an instrument for the training of students in the area of satellites, with this the young students are made aware that the development of satellites in Mexico is possible. The development of micro and nanosatellites represents Mexico's opportunity in space because the development of large satellites worth millions of dollars is practically impossible. The economy related to satellite services has evolved by leaps and bounds, and today, large companies and leading countries in the space sector are investing in prominent constellations of Nanosatellites.

The KuauhtliSAT nanosatellite has a tube shape (TubeSat), and the concept arose with the company Inter Orbital (Fig.28). The primary mission of the

Fig. 26 Students integrating the UNAMSAT Microsatellite. *Source* own Research

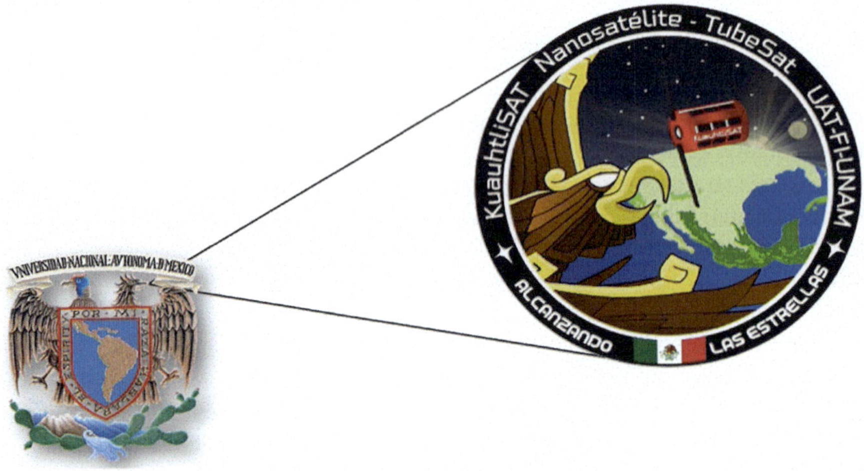

Fig. 27 University Nanosatellite KuauhtliSAT. *Source* Own Research

nanosatellite is to demonstrate its technology in space, train human resources in space area, and perform remote sensing functions. The nanosatellite comprises side panels and sections of space aluminum to give rigidity to the satellite while helping to dissipate heat. KuauhtliSat is made up of various subsystems such as the power system, on-board computer (OBC), radio communications in the UHF band (Ultra High Frequency) for TT&C (Tracking, Telemetry and Command) tasks, two low-resolution COTS (Commercial Off-The-Shelf) type cameras to carry out remote sensing tasks (Fig. 28).

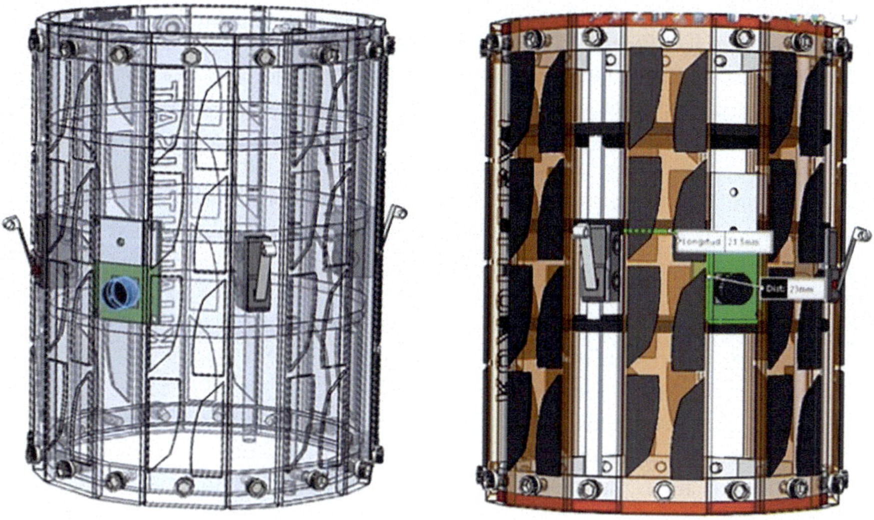

Fig. 28 Configuration of the Nanosatellite KuauhtliSat. *Source* own Research

The KuauhtliSat project involves students improving the designs of the different subsystems that make up the nanosatellite increasingly. Students carry out simulations with the help of specialized software, prototypes, and tests to reach the final design of the nanosatellite (Fig. 29).

The energy system comprises solar panels whose main element is the Trisol-X brand triangle-shaped cells with acceptable efficiency. These cells are obtained

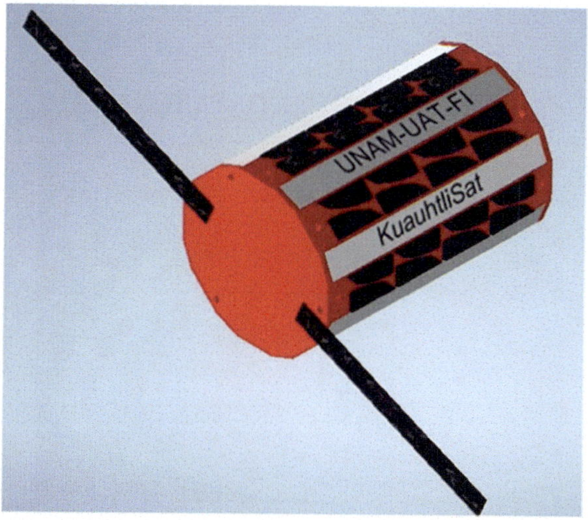

Fig. 29 Artist impression of KuauhtliSAT. *Source* own Research

from cuts of Azurspace space-grade solar cells. Made of Germanium (Ge) substrate, TrisolX cells are triple-junction GaInP/GaAs/Ge and have an efficiency of 28%. The TubeSat-type nanosatellite has eight available faces where the solar panels can be placed, and its dimensions are 127 mm high by 21.08 mm wide (Fig. 30).

When the project begins, developing each of the nanosatellite subsystems involves technological challenges. For example, a flat-sat model must be developed, followed by an engineering model, and finally, the flight model (Fig. 31).

The development of the KuauhtliSAT nanosatellite subsystems must be validated on the ground in simulated conditions that they will face in space. Sub-orbital flights must be used to validate the correct functioning of the different nanosatellite subsystems since they offer an excellent alternative to testing and validating the proper functioning of each subsystem (Fig. 32).

Given that Mexico needs to develop its satellites with Mexican engineers, it is also necessary to consider Earth observation satellites, not just telecommunications. So, the High Technology Unit has also begun including small cameras on

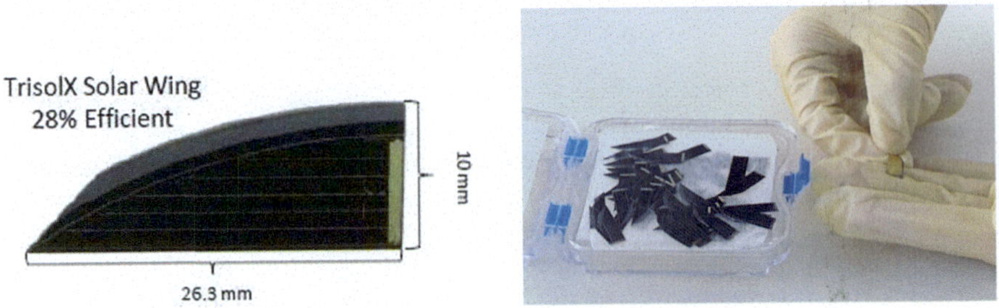

Fig. 30 Solar cells used for KuauhtliSAT. *Source* own Research

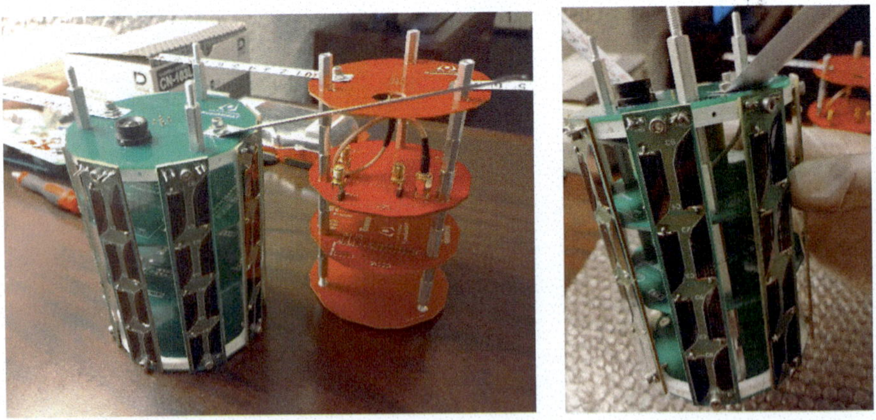

Fig. 31 Engineering model of KuauhtliSAT with solar panels. *Source* own Research

Fig. 32 Images obtained with KuauhtliSat cameras during a suborbital flight. *Source* own Research

its TubeSat and CubeSat-type satellite platforms. The cameras are commercial and low-cost.

Development of the Cubesat K'OTO Standard Nanosatellite

With the arrival of nanosatellites to the High Technology Unit with KuauhtliSat, other nanosatellite initiatives were derived to work in parallel on the development of a satellite within the CubeSat standard and continue training space engineers from different universities in the state of Querétaro in Mexico. The CubeSat project's mission is to photograph the Mexican territory in the low and medium-resolution visible light spectrum and serve as a demonstrator of Mexican technology in space (Fig. 33).

Fig. 33 Projected structure of Mexican nanosatellite K'OTO. *Source* own Research

Fig. 34 UHF Radio Communications Card for the K'OTO CubeSat. *Source* own Research

The CubeSat K'OTO project has served as a linking instrument with national and Latin American companies to realize some components. The antenna, for example, in the S-band, was initially designed and developed by some UNAM Faculty of Engineering students and was channeled to a company that adjusted the design and finally manufactured the antenna to be used in space. Likewise, the orientation system was developed by academics and students from the High Technology Unit of the Faculty of Engineering of the UNAM. Currently, the project is still being developed and tested. It is expected that by 2025, it will reach the International Space Station - ISS from where it will be released to reach an orbit approximately 400 km high and travel at close to eight kilometers per second, which is at the speed at which the ISS travels (Fig. 34).

In the 2000s, another Mexican project was born under the CubeSat standard, AzTechSat-1. The project was carried out in collaboration with NASA and the Popular Autonomous University of Puebla (UPAEP). It was named that way to give it an identity linked to Mexican culture. The project adds to Mexico's efforts to develop satellites and space instrumentation by integrating an interdisciplinary team of engineering students from the UPAEP (Fig. 35).[31]

The primary mission of the AzTechSat-1 project was to demonstrate satellite communication with the GlobalStar constellation and to demonstrate that it is possible to expand the security of radio communications using commercial satellite constellations. The success of this mission would open the door for communications from other Mexican nanosatellite projects in future missions.

UNAM launched the first microsatellite into orbit in 1996, and again, another university, with advice from NASA, integrated another CubeSat-type satellite.

[31] Milenio, "Primer Nanosatélite mexicano Aztechsat-1 cumple con éxito misión en el espacio", 30 january 2025, https://www.milenio.com/negocios/nanosatelite-mexicano-aztechsat-1-exitosa-cumple-mision-exito.

Fig. 35 CubeSat AzTechSat nanosatellite-1

Fig. 36 Deployment of the AzTechSat-1 Nanosatellite from the ISS (NASA)

AzTechSat-1 successfully entered orbit in 2020, making it the first Mexican nanosatellite launched into space from the International Space Station (Fig. 36).[32]

7 Conclusions

Mexico has been involved in space-related activities since the 1950s. At that time, the first experimental rockets, including the SCT1 and SCT2, were launched, propelled by personnel from the Ministry of Communications and Transportation. Later, Engineer Walter Cross Buchanan decisively promoted these activities to develop rocketry in our country.

[32] Gob," Nanosatélite mexicano Aztechsat-1 inicia su misión en el espacio", 30 january 2025, https://www.gob.mx/aem/prensa/nanosatelite-mexicano-aztechsat-1-inicia-su-mision-en-el-espacio-235269?idiom=es.

The National Commission for Outer Space (CNEE) was created by presidential decree in 1962. It is considered the predecessor of the current Mexican Space Agency. The CNEE is a dependency of the Ministry of Communications and Transportation, whose purpose is to control and promote everything related to research, exploitation, and the use of outer space for peaceful purposes in our country. The programs the CNEE wanted to develop included sounding rockets, sounding balloons, meteorological satellites, remote sensing, basic and applied research, and space law. Unfortunately, as has happened throughout our country's history, in 1977, the CNEE was also extinguished by presidential decree, thus losing an enormous opportunity for Mexico to be a world power in the research and development of space activities.

As described above, Mexico has had four generations of commercial geostationary communications satellites, none of which have used its indigenous technology. Since the conception of the first generation of Mexican satellites, the Morelos Satellite System, in the early 1980s, there has been a total lack of a space policy by the Mexican State, a government failure to privilege social problems over the interests of large companies, who have set the course of public policies in our country. As a result of this apathy shown by government agencies, the participation of Mexican engineers and scientists has been limited only to understanding and operating the terrestrial infrastructure related to the control and operation of the different available satellite systems.

Although the official discourse mentions that the acquisition of satellites with foreign technology contributes to national sovereignty, everything seems to indicate that Mexico's technological dependence on large international consortiums is becoming more and more pronounced since its technology is not 100% open to technicians and engineers who participate in this critical area at a global level. Space technology is marking the course of the economy, technology, academia, and society as a whole at a worldwide level, so it is necessary to have a well-defined space policy that encourages all sectors of society and the economy to participate in it to generate knowledge and its application in space issues, to develop the technology that allows us to achieve national autonomy sovereignty and prosperity.

It is important to highlight the great efforts that academic and research institutions have been making to generate knowledge in the space field. National institutions such as UNAM, IPN, INAOE, CICESE, and UPAEP, among others, continue to invest time and effort in the design and construction of nano satellites that will mark the course of our country in space matters in the short term.

Mexico is experiencing a new opportunity in the space area; there is already a base space infrastructure, new space projects, specialists, and excellent collaboration with universities, companies, engineers, and international researchers. Let us continue to reach the stars for the benefit of Mexico and humanity.

Jose Alberto Ramirez Aguilar Ph.D. is an associate professor at the National Autonomous University of Mexico School of Engineering. He received his Ph.D. in Technical Sciences in Radio receivers and microsatellites from the Moscow Aviation Institute—MAI, Russian Federation. He is the head of the Aerospace Engineering Department responsible for the Earth Station Laboratory and Vice-chair of the GRULAC of the International Astronautical Federation—IAF. His current research areas are Radio Frequency and Microwave Systems, GNSS, Antennas, TT&C, Nano, and Microsatellites. Likewise, in 2020, he was selected for the first Latin American manned space mission, ESAA-01EX SOMINUS AD ASTRA.

Dulce Carolina Sánchez-Hernández M.Eng. obtained the title of Communications and Electronics Engineer from UNAM and collaborated as an assistant professor of the subject. She worked at Philips Mexicana in the Consumer Electronics area as a supervisor of service centers nationwide. She worked for the television station TV AZTECA as a Maintenance Supervisor of the National Network of repeater stations and later as a Project Manager of the National Network. She traveled to the Russian Federation, where she studied for a master's degree in Mobile System Engineering at the Faculty of Radiocommunications of the Moscow Aviation Institute with honors. She currently serves as Coordinator of the Telecommunications and Network Engineering career at the Autonomous University of Querétaro.

Raúl Infante-Galindo M.Sc. obtained a degree in Communications and Electronics Engineering from the Escuela Superior de Ingeniería Mecánica y Eléctrica of the IPN. He later graduated a Master of Science at the Center for Scientific Research and Higher Education of Ensenada (CICESE) with a specialization in High-Frequency Electronics. For 9 years, he worked in the company Satélites Mexicanos, an operator of SATMEX satellites, where he worked in the area of Radio Frequency to control satellites. Currently, he is the coordinator of the Telecommunications and Network Engineering course at the Faculty of Computer Science of the Autonomous University of Querétaro.

Dr. Jorge Alfredo Ferrer-Pérez is an associate professor at the National University Autonomous of Mexico School of Engineering. He received his Ph.D. in Aerospace and Mechanical Engineering from the University of Notre Dame, South Bend, in the United States. He is part of the Aerospace Engineering Department and is responsible for the Space Propulsion and Thermo-vacuum lab. This facility belongs to the Space and Automotive Engineering National Laboratory located at Juriquilla. His current research areas are nano-heat transfer in solid-state devices, thermal control, space propulsion, small satellites, and space technology development.

Spaceflight Environmental Test Facilities at UNAM to Develop Space Projects

Jorge Alfredo Ferrer-Pérez, Carlos Romo-Fuentes, Rafael Guadalupe Chávez-Moreno, José Alberto Ramírez-Aguilar, Bryanda Guadalupe Reyes-Tesillo, Alfredo Rivas-Iturbe, Bárbara Bemudez-Reyes, and Dafne Gaviria-Arcila

Abstract

A satellite or platform is an artifact of electronic circuits and mechanical systems that orbit around the Earth and other celestial bodies. Satellites can be used for communications services, weather prediction and monitoring, Earth observation, navigation, military applications such as national security, and scientific missions. A satellite is made up of subsystems with specific functions that impact its operation, including the mechanical structure subsystem, propulsion, thermal control, power supply, telemetry-tracking-and-command (TT&C), altitude and orbit control, onboard computer-and-data-handling (OC&DH), and most importantly, the instrumentation corresponding to the payload. This

J. A. Ferrer-Pérez (✉) · C. Romo-Fuentes · R. G. Chávez-Moreno · J. A. Ramírez-Aguilar · A. Rivas-Iturbe · B. Bemudez-Reyes · D. Gaviria-Arcila
Advanced Technology Unit, School of Engineering, UNAM, Juriquilla Querétaro, Mexico
e-mail: ferrerp@unam.mx

C. Romo-Fuentes
e-mail: carlosrf@unam.mx

R. G. Chávez-Moreno
e-mail: rchavez@comunidad.unam.mx

J. A. Ramírez-Aguilar
e-mail: albert09@unam.mx

B. Bemudez-Reyes
e-mail: barbara.bermudez@ingenieria.unam.edu

D. Gaviria-Arcila
e-mail: dafne.gaviria@comunidad.unam.mx

B. G. Reyes-Tesillo
Centro de Física Aplicada y Tecnología Avanzada, PCeIM, UNAM, Juriquilla Querétaro, Mexico
e-mail: b.reyes@comunidad.unam.mx

J. A. Ferrer Pérez and A. Froehlich (eds.), *Space Fostering Mexican Society*, Southern Space Studies, https://doi.org/10.1007/978-3-031-92617-4_7

manuscript summarizes on one hand the required space-relevant environmental tests to validate spacecraft functionality on Earth before being deployed in space. On the other hand it analyses the facilities in Mexico to develop space technology at UNAM, such as the National Laboratory of Space and Automotive Engineering (LN-INGEA).

1 Introduction

In June 1985, the first Mexican satellite was launched, called Morelos I. Originally, Morelos I was operated in the geostationary orbital position 113.5° West. However, its replacement was later placed half a degree to the East, positioning it at 113.0° West, which Mexico currently occupies. The Morelos II satellite, the second of this first generation, was placed in its nominal orbital position at 116.8° West, beginning its operations on 1 September 1989.[1] On Friday, 2 October 2015, the Morelos 3 satellite was launched from Space Launch Complex-41 at Cape Canaveral Air Force Station, Florida, by ULA. The launch window was 6:08–6:28 a.m. EDT.[2] Figure 1 shows the mission profile and ground trace of Morelos-3.[3]

As expected, the conditions of the satellites during their journey from Earth to space are diverse and extreme. On Earth, the satellite has to withstand severe mechanical, acoustic, and vibration conditions caused by the rocket's movements during launch. When the satellite separates from the main rocket body, the necessary mechanisms must be in place to ensure the proper deployment of the satellite and various components, such as antennas and solar panels, that allow the satellite to operate. Finally, once the satellite is in space, it must withstand all the conditions of the space environment: high-energy particles, ultra-high vacuum conditions, extreme temperatures, space debris, micro-meteorites, and dust particles, generating an aggressive and dangerous environment. To reduce the risk of spacecraft failure, tests are performed in specialized facilities worldwide to reproduce different environmental conditions.

Some global and local satellite numbers are listed by Bryce Tech. According to the evaluations and numbers available, in the third quarter of 2024 (Q3), Space X alone performed 27 launches of 57 worldwide. Figure 2 shows these launches by country. Regarding spacecraft, again, Space X led global efforts to place 518

[1] Government of Mexico, "Estudio del Desarrollo de los Satélites Mexicanos y constelaciones de órbitas bajas", AEM, 21 January 2025, https://www.gob.mx/cms/uploads/attachment/file/638394/Estudio_del_Desarrollo_de_los_Satelites_Mexicanos_y_Constelaciones_de_Orbitas_Bajas.pdf (all websites cited in this publication were last accessed and verified on 30 January 2025).

[2] ULA, "Atlas V to Launch Morelos-3", 20 January 2025, https://www.ulalaunch.com/about/news/2015/10/02/atlas-v-to-launch-morelos-3.

[3] ULA, "Atlas V Morelos-3 Mission Booklet", 20 January 2025, https://www.ulalaunch.com/docs/default-source/news-items/av_morelos3_mob.pdf.

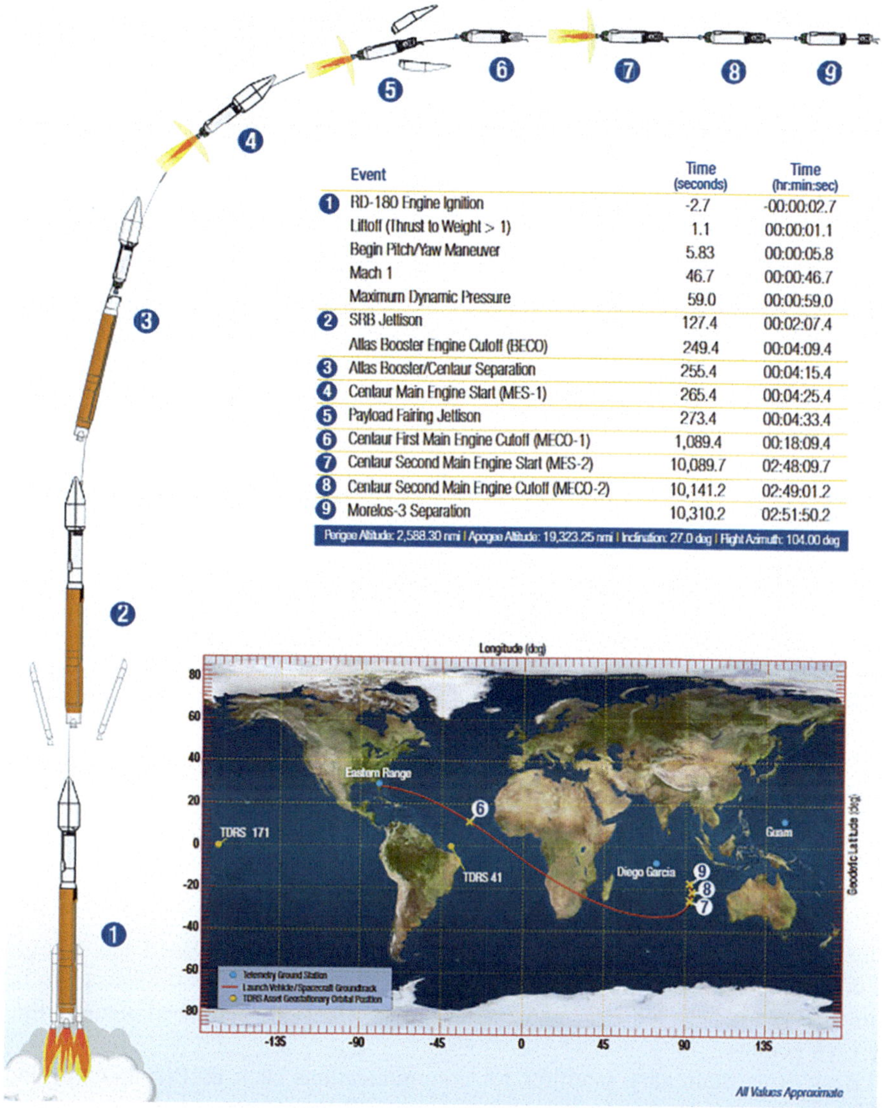

Fig. 1 Mission profile and ground trace of Morelos-3

Fig. 2 Launched satellite by country and Europe during Q3 2024

spacecraft in orbit.[4] In terms of small satellites, a total of 2.860 spacecraft were launched in the year 2023.[5]

For the year 2024 the numbers are as follows[6]:

- Nanosats launched (2.806),
- CubeSats launched (2.596),
- PocketQubes launched (83) and
- Interplanetary CubeSats (18).

Other aspects are: Most nanosats on a single rocket were 120 and the number of countries with nanosats was 89. A forecast estimates over 2.080 nanosats to be launched yearly between 2022 and 2027. Figure 3 shows the launched nanosatellite by each country.

Besides geostationary satellites for communications such as Bicentenario and Morenos-3,[7] México has launched five nanosatellites of different sizes and purposes: Painani-I, Painani-II, AztechSat-1, NanoConnect-2, and D2/Atlacom-1.

---

[4] Brycetech, "Global Orbital Space Launches Q3 2024", 20 January 2025, https://brycetech.com/reports/report-documents/Bryce_Briefing_2024_Q3.pdf,

[5] Brycetech, "Smallsats by the Numbers 2024", 20 January 2025, https://brycetech.com/reports/report-documents/Bryce_Smallsats_2024.pdf,

[6] Nanosats Database, 20 January 2025, https://www.nanosats.eu/.

[7] Ramirez-Aguilar, J.A., Sanchez-Hernandez, D.C., Ferrer-Perez, J., Chávez-Moreno, R., Romo-Fuentes, C. Satellites for the Benefit of Mexican People: Past, Present and Future. In: Froehlich,

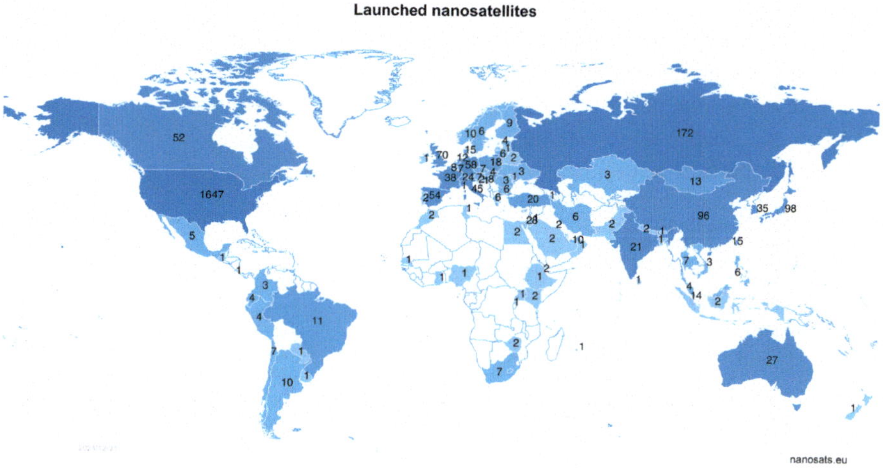

Fig. 3 Launched nanosatellite by country up to 31 December 2024

Further projects are undergoing, such as K'OTO and IXAYA, and others were announced by the Current Mexican Government (2024–2030), such as a new geostationary satellite system.[8]

2 Spaceflight Environmental Tests

All the spacecraft launched so far need to pass several tests on Earth to reproduce the changing environment along the path from Earth to space. These tests aim to analyze the spacecraft's performance, subsystems, or components in different environmental conditions reproduced in specialized facilities. In general, the possible that are performed in practice are[9]:

a. Electrical interface test
b. Functional test
c. Mission test
d. Total Ionization Dose (TID) test

A. (eds) Space Fostering Latin American Societies. Southern Space Studies. Springer, Cham. 2021. https://doi.org/10.1007/978-3-030-73287-5_2.

[8] Government of México, "Presidenta Claudia Sheinbaum presenta la nueva Agencia de Transformación Digital para simplificar trámites y evitar la corrupción," 20 January 2025, https://www.gob.mx/presidencia/prensa/presidenta-claudia-sheinbaum-presenta-la-nueva-agencia-de-transformacion-digital-para-simplificar-tramites-y-evitar-la-corrupcion.

[9] ISO, "Space systems — Design qualification and acceptance tests of small spacecraft and units," ISO 19683:2017, 20 January 2025, https://www.iso.org/es/contents/data/standard/06/60/66008.html.

e. Single Event Effects (SEE) test
f. Spacecraft Charging Induced Electrostatic Discharge (ESD) test
g. Electromagnetic Compatibility (EMC) test
h. Deployment test
i. Magnetic field test
j. Antenna pattern test
k. Launcher/Spacecraft interface test
l. Quasi-static load test
m. Sinusoidal vibration test
n. Random vibration test
o. Acoustic test
p. Shock test
q. Thermal balance test
r. Thermal vacuum test
s. Functional test in vacuum
t. Cold/Hot start test
u. Thermal cycle functional test
v. Thermal cycle endurance test
w. Pressure test
x. Leakage test
y. Microvibration test
z. Burn-in and wear-in test
aa. Bake out and outgas test.

Each of these individual tests can be carried out with the following scopes[10]:

i. **Development Tests**. Tests are conducted on representative articles[11] to characterize engineering parameters, gather data, and validate the design approach.
ii. **Qualification Tests**. Tests are conducted to demonstrate satisfaction with design requirements, including margin and product robustness, for designs with no documented history.
iii. **Protoqualification Tests**. Tests are conducted to demonstrate satisfaction of design requirements using reduced amplitude and duration margins.
iv. **Acceptance Tests**. Vehicle, subsystem, and unit tests[12] are conducted to detect workmanship defects, demonstrate specified functional and performance requirements, and ensure the hardware is acceptable for delivery.

[10] NASA, "Space and Missile Systems Center Estándar: Test Requirements for Launch, Upper-Stage and Space Vehicles," SMC Standard SMC-S-016, 20 January 2025, https://explorers.larc.nasa.gov/2019APSMEX/MO/pdf_files/SMC-S-016_05SEP2014.pdf.
[11] Test article is a vehicle, subsystem, or unit dedicated to provide design requirement information.
[12] Lowest level of hardware assembly for which acceptance and qualification tests are required.

v. **Prelaunch Validation Tests**. Prelaunch validation tests are conducted at the launch base to demonstrate the readiness of the hardware, software, personnel procedures, and mission interfaces to support the launch and the program mission.
vi. **Post-launch Validation Tests**. Tests performed following launch to demonstrate performance, interface compatibility, calibration, and the ability to meet mission requirements.

Depending on the size of the spacecraft, only some tests can be performed to shorten the time and financial resources. For example, it is proposed to carry out only qualification acceptance and proto-flight tests for small spacecraft.[13]

2.1 Norms and Facilities

Environmental tests are executed according to norms or standards defined by national space agencies, military branches, the private sector, or international agreements. It is critical to point out that each norm is a guideline that shows what to do but does not show how to do the tests. NASA and ESA have very comprehensive open reservoirs with norms for each type of test. Likewise, SpaceX of ULA also defined the standards clients/users must fulfill to launch their spacecraft.

To implement norms, specialized facilities are needed. They consider expensive equipment and highly trained staff. They are located in space agencies (e.g., NASA, ESA, CONAE, INPE), at universities (e.g., MIT, Arizona, Surrey, Stuttgart, Luxemburg), and within private companies(e.g., Experior Labs,[14] UK Facilities,[15] Cira[16]) around the globe. They are used to test local projects or can be hired to run environmental test campaigns for clients outside their organization at a cost. It is crucial to observe the agreed standards that can be satisfied by each test entity.[17]

[13] ISO, "Space systems — Design qualification and acceptance tests of small spacecraft and units,"General Requirements, 4–5.

[14] Experior Laboratories, "Extreme environments Testing, Certification and Qualification," https://repositoriodigital.ipn.mx/bitstream/123456789/26505/1/Agenda_Estrategica_de_la_IEAE_2030_D.pdf, https://experiorlabs.com/.

[15] Science and Technology Facilities, "UK Space Facilities," 20 January 2025, https://www.ukspacefacilities.stfc.ac.uk/Pages/Industrial-Space-Test-Facilities.aspx.

[16] Cira, "Space Qualification Laboratory," 20 January 2025, https://www.cira.it/en/research-infrastructures/space-qualification-laboratory.

[17] Experior Laboratories, "Testing Standards", 21 January 2025, https://experiorlabs.com/testing-standards/.

2.2 Test Facilities in Mexico

According to the Strategic Agenda of Higher Education Institutions for the Aeronautical and Space Sectors 2030,[18] Mexico lacks national owned facilities focused on performing spaceflight environmental tests. The following universities have devoted test facilities: UNAM[19] (National University Autonomous of Mexico), IPN[20] (National Polytechnic Institute), and UPAEP[21] (Popular Autonomous University of the State of Puebla). Moreover, the Regional Space Development Center (CREDES) of the Mexican Space Agency (AEM) in the State of México is a facility projected to perform spaceflight and rocket tests in the midterm. Another CREDES dedicated to developing communication systems is in the State of Zacatecas.

2.3 Test Facilities at UNAM

UNAM has test facilities on the main campus (C.U.) in the City of México and the Juriquilla Campus in the State of Queretaro. In the C.U., there is the Laboratory of Electronic Instrumentation of Space Systems (LIESE[22]) at the School of Engineering and the Space Instrumentation Laboratory (LINX[23]) at the Nuclear Sciences Institute. At the Juriquilla Campus, the National Laboratory of Space and Automotive Engineering (LN-INGEA) is located.

LN-INGEA[24] was created in 2014 based on the analysis of the need to design, manufacture, integrate, and test space and automotive systems in the country. In this way, it supports local technological development (Querétaro) and gradually expands its capabilities nationwide. Currently, LN-INGEA is housed in the Advanced Technology Unit[25] of the School of Engineering of the UNAM and participates in different initiatives of national impact, such as the development of micro and nano-satellite platforms, generation of NMX standards, and academic

[18] SEP, "Agenda Estratégica de las Instituciones de Educación Superior para los sectores Aeronáutico y Espacial 2030", August 2022, https://repositoriodigital.ipn.mx/bitstream/123456789/26505/1/Agenda_Estrategica_de_la_IEAE_2030_D.pdf.

[19] UNAM, "Programa Espacial Universitario," 21 January 2025, https://peu.unam.mx/index.html.

[20] IPN, "Centro de Desarrollo Aeroespacial," 21 January 2025, https://www.ipn.mx/cda/.

[21] UPAEP, "Ingeniería Aeroespacial," 21 January 2025, https://upaep.mx/licenciaturas/ingenieria-aeroespacial.

[22] FI-UNAM, "Laboratorio de Instrumentación Electrónica de Sistemas Espaciales," 21 January 2025, https://odin.fi-b.unam.mx/liese/.

[23] ICN-UNAM, "Laboratorio de Instrumentación Espacial," 21 January 2025, https://epistemia.nucleares.unam.mx/web?name=linx.

[24] LNINGEA, "National Laboratory of Space and Automotive Engineering," 21 August, 2022, https://www.ingenieria.unam.mx/uat/lniea.php.

[25] FI-UAT, "Advance Technology Unit of the Faculty of Engineering," 21 August 2022, https://www.ingenieria.unam.mx/uat/.

undergraduate programs in Aerospace Engineering.[26,27] Additionally, the members of the LN-INGEA working group generate patents, scientific articles, book chapters, conferences, as well as undergraduate, masters and doctoral theses as part of the training of undergraduate and graduate human resources that allow detonating new areas of knowledge for the strengthening of the technological independence of the nation in key sectors.

At the Federal level, LN-INGEA contributes to strategy 2.2, objective 2 of the National Space Activities Program 2020–2024,[28] published in the Official Gazette of the Federation, to the lines of action: 4.A.1, 4.B.1 and 4.C.1. It is necessary to mention that the activities developed in LN-INGEA are aligned not only with the national development plan of the federal government but also with the Sustainable Development Goals of the United Nations, which are the reference for the Strategic National Programs (PRONACES).[29]

From a regional perspective, the LN-INGEA aims to support higher education institutions, government agencies, and private initiatives in the country's central region to design, manufacture, integrate, and test space and automotive systems. The Querétaro cluster, where the LN-INGEA is located, occupies third place in importance at the national level regarding the number of companies related to the development of aerospace technology and is first in investment. Additionally, the government of Querétaro, in its strategic plan, recognizes the need to strengthen the aerospace and automotive areas in the years to come.

In addition, LN-INGEA is linked to the thematic networks of National Council of Science and Technology (CONACyT), which allows supporting the development of projects within the framework of the work plan of the Space Science and Technology Network (REDCyTE), where the participating members have close collaboration and use of the facilities to promote the space area. On the other hand, LN-INGEA participates with other national laboratories in the country, such as the National Space Climate Laboratory (LANCE) and the National Earth Observation Laboratory (LANOT), in the generation of projects with high social and technological impact, allowing the sharing of infrastructure and the proposal of innovative solutions to the current national problems of the country.

The following sections provide a general overview of the laboratories dedicated to developing space systems at LN-INGEA and mention some of the projects currently being carried out.

[26] FI-UNAM, "Bachelor of Aerospace Engineering," UNAM, 31 March 2023, https://www.ingenieria.unam.mx/programas_academicos/licenciatura/aeroespacial.php.
[27] ENESJ-UNAM, "Bachelor of Aerospace Engineering," 31 March 2023, https://www.enesjuriquilla.unam.mx/?page_id=6169#.
[28] AEM, "National Space Activities Program 2020–2024", 21 July 2022, https://www.gob.mx/aem/documentos/programa-nacional-de-actividades-espaciales-2020-2024.
[29] National Council of Science and Technology, "Strategic National Programs," CONACyT, 4 April 2023, https://conacyt.mx/pronaces/.

3 LN-INGEA Laboratories

As stated above, specialized infrastructure is required to validate the systems manufactured on the ground under space conditions to develop space systems such as nanosatellites. These tests are the bridge that allows the development of technology that survives the environmental conditions to which a satellite is subject from its launch to its entry into orbit. Although various higher education institutions nationwide have different infrastructures to develop this type of system, not all have all the laboratories and equipment required to design, manufacture, test, and operate space systems.

LN-INGEA has the following laboratories dedicated to developing space systems:

1. Vibrations,
2. Space propulsion and thermal vacuum,
3. ISO 5 clean room,
4. Electromagnetic compatibility,
5. Printed circuits and embedded systems,
6. Integration and training of space systems and
7. Ground station.

3.1 Vibration Laboratory

This laboratory has three vibration tables of different capacities, acceleration sensors, and a Polytec 500 laser Vibrometer to measure the deformation of a sample in real-time without contact, which represents a significant advantage over traditional measurements. Figure 4 shows this equipment.[30]

3.2 Space Propulsion and Thermovacuum Laboratory

The Space Propulsion and Thermovacuum Laboratory (LPET)[31] has different equipment for performing vacuum and thermal vacuum tests. Its main characteristics are:

- Cylindrical vacuum chamber dimensions: 2 m. diameter × 3.5 m. length,
- Minimum vacuum target ~ 10E-4 Torr,
- Temperature working range from − 100 °C to + 120 °C,
- Maximum sample dimensions of 1 m × 1 m × 1 m,

[30] Polytec, "Polytec Scanning Vibrometer", 29 january 2025, https://www.polytec.com/int/vibrometry/products/full-field-vibrometers/psv-500-scanning-vibrometer.
[31] LPET, "Space Propulsion and Thermovacuum Laboratory," 31 March 31 2023, https://www.ingenieria.unam.mx/uat/LabTermoVacio.php.

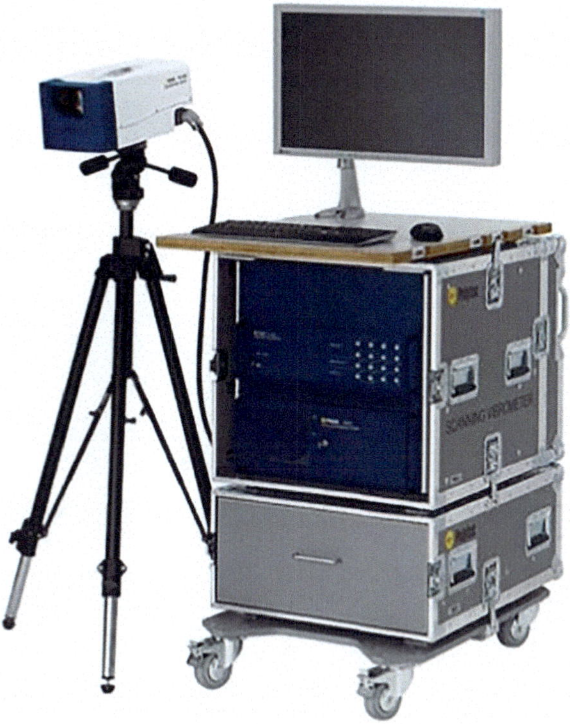

Fig. 4 Polytec 500 laser Vibrometer

- Unique equipment in the nation to perform thermal-vacuum tests of components, subsystems, and complete systems,
- It allows research and development of space thrusters for small satellites,
- It can be used to perform outgassing tests (Fig. 5).

The following elements form LPET:

- One mechanical vacuum pump,
- One turbomolecular vacuum pump,
- Heating system (lamps),
- Cooling system (cryogenic chillers with copper coils),
- Control system.

Figure 6a presents the above elements, and Fig. 6b visualizes the cooling and heating systems.

Finally, it is relevant to note that on 22 April 2022, patent title 392148 was granted to the invention HALL EFFECT THRUSTER FOR SPACE VEHICLES

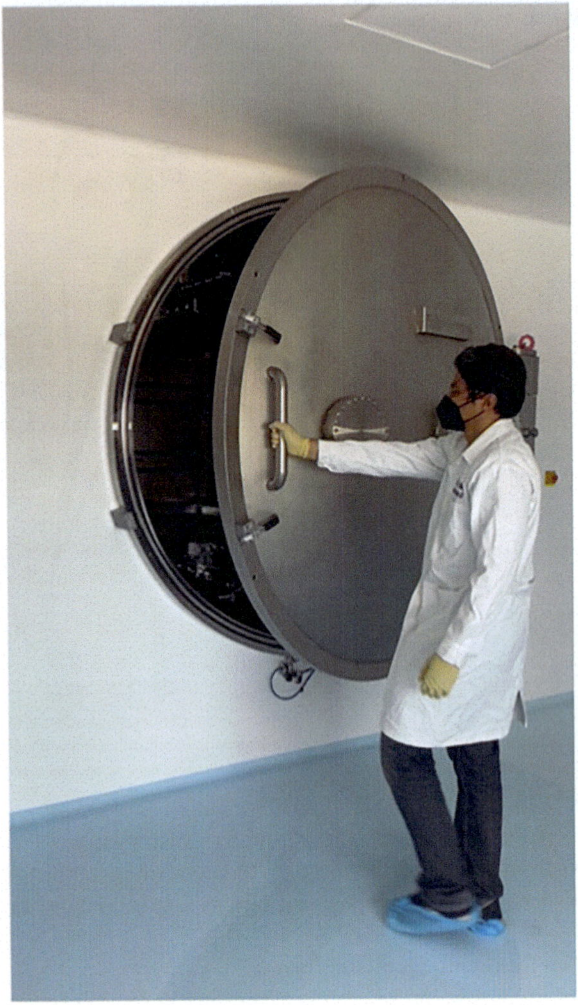

Fig. 5 Thermal vacuum chamber door. Courtesy of LN-INGEA

for 20 years,[32] which represents the first national effort to develop proprietary propulsion technology in México.

A thermal vacuum (TV) preliminary test campaign was conducted to test the LPET equipment. The test sample was a cubic mock model with dimensions similar to a 1U CubeSat. The objective was to reproduce the TV test required by

[32] Mexican Institute of Industrial Protection, "Propulsor de Efecto Hall para Vehículos Espaciales MX/a/2018/009345", IMPI, 31 March 2023, https://vidoc.impi.gob.mx/visor?usr=SIGA&texp=SI&tdoc=E&id=MX/a/2018/009345.

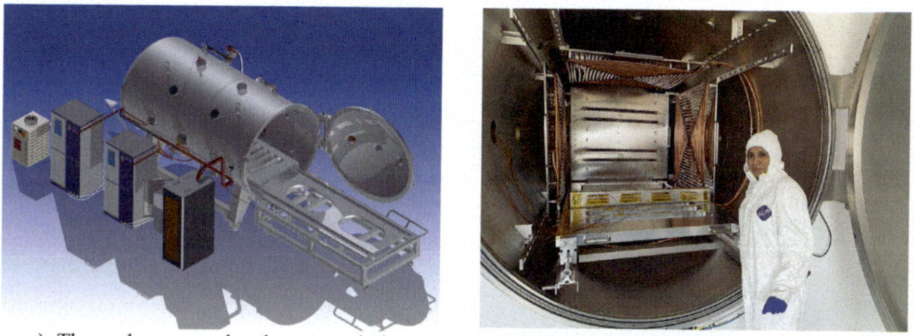

a). Thermal vacuum chamber general view b). Cooling and heating thermal vacuum systems

Fig. 6 Thermal vacuum chamber general and inner view. Courtesy of LN-INGEA

JAXA in the JEM Payload Accommodation Handbook,[33] which states that outside the ISS, the temperature range is − 15 to + 60 [°C] (when the satellite is in the J-SSOD). Moreover, the chamber pressure was selected as the order of 10^{-4} Torr as stated SMC-S-016. Nevertheless, this pressure can differ depending on the required standard, for example, 1×10^{-3} Pa (~ 7.5×10^{-6} Torr) according to ISO 19683:2017. The difference between this selection impacts the vacuum system (i.e., vacuum pump system).

A sarcophagus was created and suspended within the chamber to improve the sample heating and cooling process, as shown in Fig. 7a. The dimensions of this volume are 16 cm × 16 cm × 40 cm. Figure 7b shows the position of the sample inside the sarcophagus, and Fig. 7c presents its location inside the chamber. It is essential to mention that the distance from the heating lamps is 1.3 m, and two thermocouples (type T) were set in positions s6 and s12. With this arrangement, the environmental temperature can be estimated. The overall test consisted of two temperature cycles: the first went from + 60 °C to – 15 °C. Meanwhile, the second one went from + 40 °C to – 20 °C.

Figure 8 presents the pressure profile along all the TV tests. From this figure, it is possible to observe that 10^{-4} was achieved before the first cycle started. When the temperature begins to rise, the pressure also increases as expected. The pressure decreases as the temperature decreases, reaching 10^{-6} Torr vacuum levels.

Figure 9 shows the temperature profiles of sensors s6 and s12. Once the target temperatures were reached, they remained for approximately one hour. The transition time between the heating and cooling conditions was around 2.6 h, and the heating slope was 4.5 °C/min. This slope can be controlled for the heating process. Nevertheless, the time the cooling process takes to decrease temperature is determined by the Cryptogenic chillers used, and it cannot be tuned.

[33] JAXA, "JEM Payload Accommodation Handbook JX-ESPC-101133-E," 22 January 2025, https://humans-in-space.jaxa.jp/kibouser/library/item/jx-espc_8e_en.pdf.

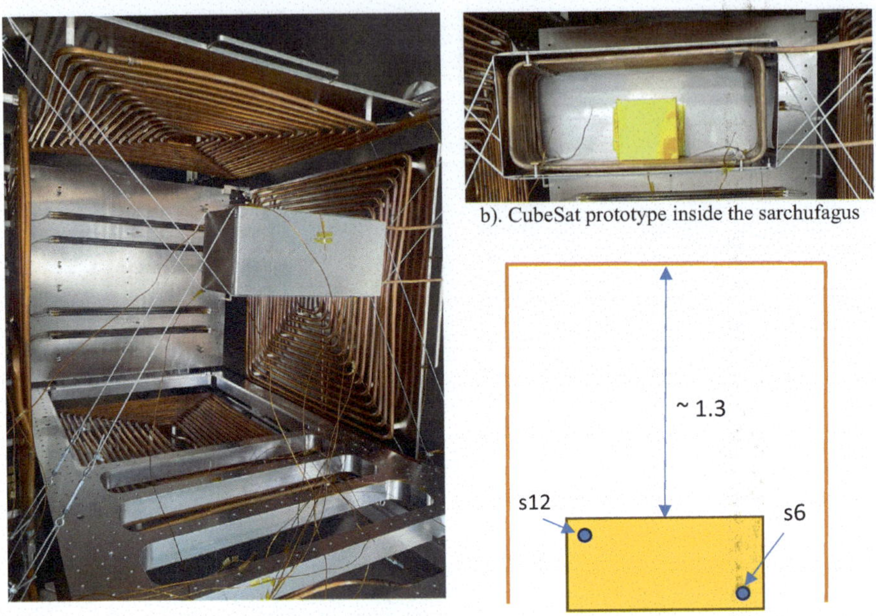

a). Inner sarcophagus suspended inside the thermal vacuum chamber.

b). CubeSat prototype inside the sarchufagus

c). Sarcufagus position.

Fig. 7 Thermal vacuum chamber inner sarcophagus. Courtesy of LN-INGEA

3.3 ISO 5 Clean Room

The LN-INGEA has an ISO 5 clean room with access to the LPET vacuum chamber inlet. This kind of rooms are critical for satellite integration with extremely sensitive optical systems.[34] The clean room has a laminar flow area on a table for integrating space systems (Fig. 10).

3.4 Laboratory of Electromagnetic Compatibility

This Laboratory of Electromagnetic Compatibility (EMC) facility consists of a rectangular-shaped laboratory with the following dimensions of length (L), width (W), and height (H)[35]:

- Internal dimensions: 7.0 m × 6.7 m × 4.3 m,

[34] NASA, "Cleanroom Tech Key to Success in Space", 10 September of 2021, https://www.nasa.gov/organizations/cleanroom-tech-key-to-success-in-space/

[35] EMC Lab, "Electromagnetic Compatibility Laboratory," 31 March 2023, https://www.ingenieria.unam.mx/uat/LabCompatibilidadElectromagnetica.php.

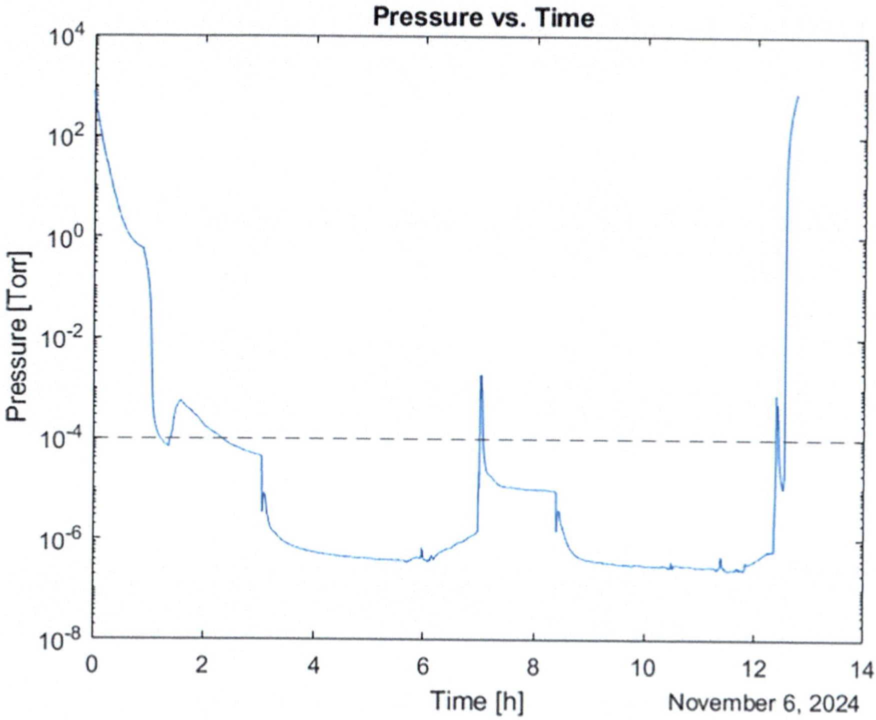

Fig. 8 Pressure profile. Courtesy of LN-INGEA

- External dimensions: 7.3 m × 7.0 m × 4.8 m and
- Effective working space: 5.8 m × 4.9 m × 3.4 m.

The EMC Lab covered the following standards:

- MIL-STD 462E—military and space industry,
- RTCA-DO-160G—aeronautical industry,
- CISPR 25/ISO 11452-2/SAE J1113/21/SAE J1113/41—automotive industry,
- CISPR 14—metal-mechanical industry, household appliances.

The services that can be provided as well are:

- Characterization of radiated and conducted emissions,
- Measurement of S parameters,
- Near-field probes E and H,
- EM emissions characterization for printed circuits with EMScan equipment,
- Numerical simulations using FEKO Software (academic license),
- NSA/VSWR testing performed in the laboratory for accreditation (December 2017) (Fig. 11).

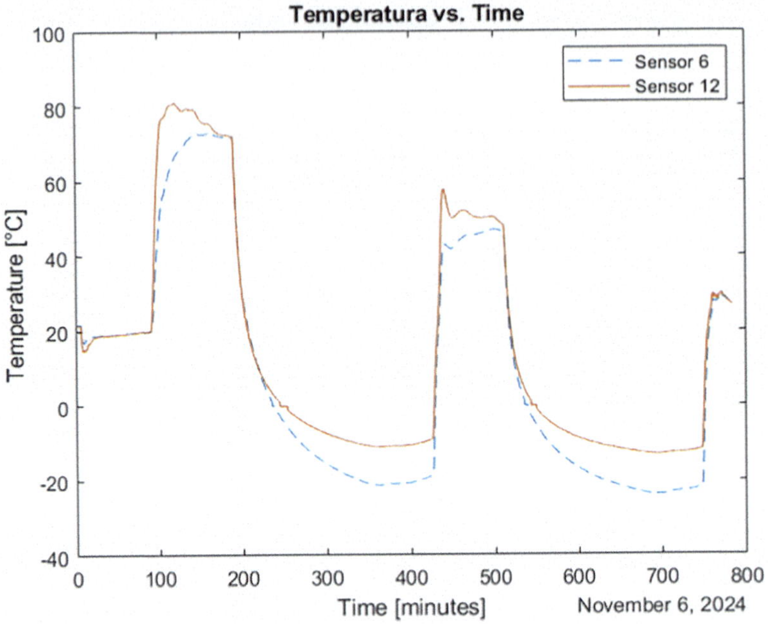

Fig. 9 Temperatures profiles. Courtesy of LN-INGEA

Fig. 10 Clean room and thermal vacuum chamber. Courtesy of LN-INGEA

Fig. 11 Electromagnetic compatibility chamber. Courtesy of LN-INGEA

On 17 November 2022, the Mexican Accreditation Entity (EMA) accredited the ECM Lab as a Testing Laboratory by the requirements established in the NMX-EC-17025-IMNC-2018 ISO/IEC 17025:2017 standard for conformity assessment activities in Electrical Electronics.

3.5 Printed Circuits and Embedded Systems Laboratory

This Printed Circuits and Embedded Systems Laboratory uses a technique for developing and testing complex real-time embedded systems.[36] For this purpose, there is an Opal-RT 5600 real-time computer with test modules for automotive and aerospace industry protocols: CAN, ARINC 429, MIL-STD 1553, Intel Xeon E5 multi-core processor, 3.2 GHz.

The inputs and outputs of the computer Opal-RT 5600 are:

- Digital: 32 I/O per slot,
- Analog: 16 per slot.

And the external expansion modules:

- NI CompactDAQ,
- NI CompactRIO,
- PXI modules.

[36] Lab HILSIL, "Hil and Sil Laboratory," 31 March 2023, https://www.ingenieria.unam.mx/uat/LabHILSIL.php.

Fig. 12 Printed circuits and embedded systems laboratory. Courtesy of LN-INGEA

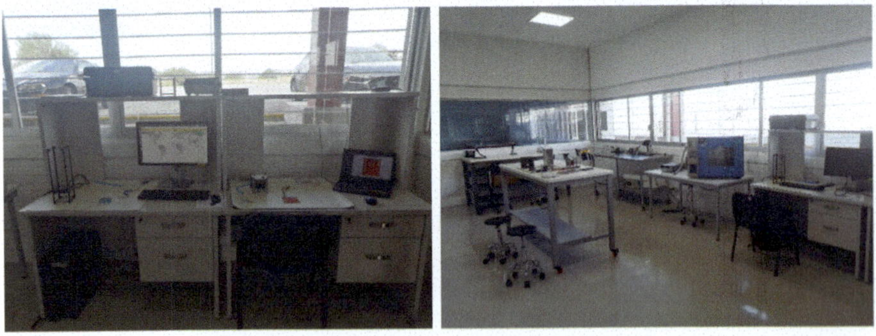

Fig. 13 Laboratory for integration and training of space systems. Courtesy of LN-INGEA

Additionally, this laboratory can manufacture PCBs of up to 24 layers to design multilayer printed circuit boards with a track resolution of over 150 μm. We use laser and UV photolytic processes and a 300 mm $\times$ 400 mm work surface (Fig. 12).

Additionally, there is equipment for performing drilling and cutting using CNC, soldering using a reflow oven and local soldering methods, mounting surface-mount-components (SMD) using pick-and-place equipment, and platinum plating of perforations (PTH) using a chemical process.[37]

3.6 Space Systems Integration and Training Laboratory

This Space Systems Integration and Training Laboratory has the infrastructure to design a space mission and manufacture different subsystems. For this purpose, there are workstations, educational nanosatellite kits, soldering stations, microscopes, IR-cameras, analytical balances, a vacuum chamber, and a mass spectrometer to perform residual gas analysis due to outgassing effects (Fig. 13).

[37] Lab Sis Emb, "Embedded Systems Laboratory," 31 March 2023, https://www.ingenieria.unam.mx/uat/LabSistemasEmbebidos.php.

Fig. 14 Ground station: Dish Antenna. Courtesy of LN-INGEA

3.7 Ground Station

The Ground Station[38] has Tracking, Telemetry, and Command (TT&C) capabilities. It can receive images from nano- and microsatellites, service and payload telemetry, and command satellites placed in low Earth orbit (LEO, ~ 900 km altitude).

The station covers frequency ranges (VHF and UHF, S- and X-band) and is authorized by the IFT, ITU, and IARU.

The station offers training in the operation via radio frequency of scientific satellites, radio amateurs, and the International Space Station (ISS), as well as orientation of satellite dishes, calculations of links with GEO satellites, design of "patch"-type antennas for different applications, and development of RF and Microwave circuits (Fig. 14).

[38] Est Control Sat, "Satellite Control Station," 31 March 2023, https://www.ingenieria.unam.mx/uat/EstacionControlSatelital.php.

4 LN-INGEA Space Projects

The LN-INGEA has several lines of action related to human resources training, technology development, research, and services. In this sense, projects are a central point allowing high-impact product generation. Some of them are described below.

4.1 Nanosatellite Projects

The K'-OTO Project,[39] whose objective is to develop, integrate, and release into orbit a CubeSat-type nanosatellite from the International Space Station (ISS) to take low-medium resolution photographs of the Mexican territory and transmit them to Earth by experimenting in the amateur radio band. The development of this experimental satellite has been strongly promoted by the Secretariat of Sustainable Development (SEDESU) of the State of Querétaro and the UAT FI-UNAM. This project involves mainly undergraduate students from various institutions, such as UNAM, UAQ, UNITEC, ITQ, and UPSRJ, from multiple areas, promoting multidisciplinary work and promoting the training of human resources in the space area through teamwork, multidisciplinary and multi-institutional collaboration, promoting the use of experimentation in the amateur radio bands and at the same time contributing to the training of space engineers and the technological development of the country, with LN-INGEA being the entity in charge of carrying out the integration and pre-certification tests of this project (Fig. 15).

4.2 National Technical Committee for Space Standardization for the Creation of (Mexican Norm) NMX Standards

The National Technical Committee for Space Standardization (COTENNE) of the Ministry of Economy and the Mexican Space Agency (AEM) is responsible for developing the Mexican NMX standards project. This committee comprises federal entities, space companies, universities, and research centers. One of the standards developed was the Mexican Standard NMX-AE-001-SCFI-2018 "SPACE SYSTEMS - CUBESATS SATELLITE DESIGN - REQUIREMENTS AND CLASSIFICATION," published in the Official Gazette of the Federation.[40]

[39] UAT, "K'OTO UNIVERSITY NANOSATELLITE PROJECT," 21 August 2022, https://www.ingenieria.unam.mx/k-oto/.

[40] Official Gazette of the Federation, "DECLARATION of validity of the Mexican Standard NMX-NMX-AE-001-SCFI-2018," DOF, 22 August 2018, https://dof.gob.mx/nota_detalle.php?codigo=5535554&fecha=22/08/2018#gsc.tab=0.

Fig. 15 K'OTO nanosatellite project. Courtesy of LN-INGEA

Since 2018, the High Technology Unit that houses the LN-INGEA has actively participated in leading two NMX standard initiatives, one related to electromagnetic compatibility and another about qualification and acceptance tests for small ships.

4.3 EMC Standard

The Electromagnetic Compatibility (EMC) work group led the initiative to develop the Mexican standard project, PROY-NMX-AE-000-SCFI-2019, "Space systems - electromagnetic compatibility requirements."

This Mexican standard addresses equipment level requirements, verification, and the fundamentals of system-level compatibility care used in the development and implementation of complete space systems where requirements are included at all of the following levels:

- General system requirements,
- Specific system requirements,
- Electromagnetic interference requirements at the equipment level.

The NMX-AE-000-SCFI-2019 standard does not include detailed design requirements. Instead, it presents the engineering problems that arise during the execution of the EMC control program.

This standard's requirements may be adapted based on contractual agreements. It also addresses Electromagnetic interference (EMI) test methods at the civil equipment level to minimize cost and allow the use of standard test methods.

DOF: 27/04/2022

DECLARATORIA de vigencia de la Norma Mexicana NMX-AE-003-SCFI-2021.

Al margen un sello con el Escudo Nacional, que dice: Estados Unidos Mexicanos.- ECONOMÍA.- Secretaría de Economía.- Unidad de Normatividad, Competitividad y Competencia.- Dirección General de Normas.

DECLARATORIA DE VIGENCIA DE LA NORMA MEXICANA NMX-AE-003-SCFI-2021 "SISTEMAS ESPACIALES-REQUERIMIENTOS DE COMPATIBILIDAD ELECTROMAGNÉTICA".

La Secretaría de Economía, por conducto de la Dirección General de Normas, con fundamento en lo dispuesto por los artículos 34 fracciones II, XIII y XXXIII de la Ley Orgánica de la Administración Pública Federal; 4 de la Ley Federal de Procedimiento Administrativo; 3 fracción X, 39 fracciones III y XII, 51-A, 51-B y 54 de la Ley Federal sobre Metrología y Normalización, 45, 46 y 47 del Reglamento de la Ley Federal sobre Metrología y Normalización; Tercero y Cuarto Transitorios del Decreto por el que se expide la Ley de Infraestructura de la Calidad y se abroga la Ley Federal sobre Metrología y Normalización; y 36 fracciones I, IX y XII del Reglamento Interior de la Secretaría de Economía, publica la Declaratoria de Vigencia de la Norma Mexicana que se enlista a continuación, misma que ha sido elaborada, por el Comité Técnico de Normalización Nacional del Espacio (COTENNE) de la Secretaría de Economía, lo que se hace del conocimiento de los productores, distribuidores, consumidores y del público en general.

El texto completo del documento puede ser consultado en la Dirección General de Normas de esta Secretaría, ubicada en Calle Pachuca número 189, Piso 7, Colonia Condesa, Demarcación Territorial Cuauhtémoc, Código Postal 06140, Ciudad de México, a través de una cita gestionada al correo electrónico: dgn.industriapesada@economia.gob.mx o en el Catálogo Mexicano de Normas del Sistema Integral de Normas y Evaluación de la Conformidad (SINEC), cuya dirección electrónica es: https://www.sinec.gob.mx /SINEC/Vista/Normalizacion/BusquedaNormas.xhtml.

La Norma Mexicana NMX-AE-003-SCFI-2021, entrará en vigor 60 días naturales contados a partir del día natural inmediato siguiente de la publicación de esta Declaratoria de Vigencia en el Diario Oficial de la Federación. SINEC-20210901122135511.

Fig. 16 Declaration of validity of standard NMX-AE-003-SCFI-2021

This standard does not contain EMI test limits, so test limits must be developed based on the environment, power quality definition, and operational requirements.

The NMX-AE-000-SCFI-2019 standard was approved in 2019 for publication in the Federation's Official Gazette. The National Laboratory has significantly contributed to developing national regulations for creating space systems within the nation under the EMC criterion. Figure 16 shows the declaration of validity of this standard.[41]

4.4 Galileo Information Center for Mexico

The new Galileo Information Center[42] (GIC) is an international initiative that will be operational for 36 months (starting in 2021) and will expand the ecosystem of Galileo Information Centers in Latin America thanks to the capabilities and influence of the National Laboratory housed in the High Technology Unit—UAT of the Faculty of Engineering of the UNAM. This initiative impacts several aspects of the interconnectivity of various productive sectors such as commerce, agriculture, and defense.

The GIC is coordinated with the centers in Brazil and Chile. Its link with the European Commission and strengthens it to develop new radio navigation and global positioning applications in aeronautical, automotive, telecommunications, Internet of Things (IoT) etc.

[41] Diario Oficial de la Federación, DECLARATORIA de vigencia de la Norma Mexicana NMX-AE-003-SCFI-2021, DOF, 27 of april of 2021, https://www.dof.gob.mx/nota_detalle.php?codigo=5650296&fecha=27/04/2022#gsc.tab=0.

[42] Galileo Information Center, 31 March 2023, https://galileoic-mexico.com/es/inicio/.

This GIC contributes to the European Commission's space outreach activities to promote EU Space Programmes and foster their use in the Latin American market. It will improve the visibility of European satellite navigation in the region and help promote cooperation on Galileo and EGNOS between the EU space ecosystem and regional stakeholders at different levels.

The GIC at LN-INGEA will provide valuable insights into local GNSS markets by monitoring local and regional satellite navigation initiatives and key stakeholders. This aims to understand their specificities and needs, as well as the market potential of EGNSS, providing new opportunities for the EU space ecosystem in the region. The Centre will offer communication, promotion, and training activities tailored explicitly to regional particularities.

5 Conclusions

Spaceflight Environmental Test Facilities at UNAM to develop Space Projects

To validate the development of space systems in ground laboratories, specialized Spaceflight Environmental Test Facilities are required to reproduce the environmental conditions where the satellite system has to function and survive. In other words, testing is the bridge that allows the developing technology to survive the environmental conditions to which a space system will be subject from its launch to its placement in orbit. This infrastructure for Spaceflight Environmental Test Facilities are often expensive and restricted. In Mexico, there are some entities where tests can be performed. Nevertheless, the UNAM National Laboratory for Space and Automotive Engineering (LN-INGEA) is an entity that has unique laboratories at a national level that are destined to support the development of space systems. The infrastructure in the same building allows the integration of an assembly line to design, manufacture, integrate, and test space systems under one roof, thus improving transport logistics and optimizing costs. In this way LN-INGEA is emerging as a national and international certification center for spaceflight environmental testing under international standards. This will allow different Mexican and international actors in the aerospace sector to develop space projects with high-impact national talent, contributing to the technological development of space in Mexico. Together with the generation of NMX Standards, training of human resources, and research and development activities, LN-INGEA is positioned as a key player in the nation's space sector in the coming years.

Acknowledgements The UNAM PAPIIT 101522 Grant partially funded this work.

Jorge Alfredo Ferrer-Pérez is a National University Autonomous of Mexico-School of Engineering professor. He received his Ph.D. in Aerospace and Mechanical Engineering from the University of Notre Dame, South Bend, in the United States. He is part of the Aerospace Engineering Department and is responsible for the Space Propulsion and Thermo-vacuum lab. This facility belongs to the Space and Automotive Engineering National Laboratory located at Juriquilla. His current research areas are nano-heat transfer in solid-state devices, thermal control, space propulsion, small satellites, and space technology development.

Carlos Romo Fuentes is an associate professor at the National University Autonomous of Mexico School of Engineering. He received his Ph.D. in Technical Sciences in the Design of Space Systems, considering electromagnetic compatibility criteria from the Aviation Institute of Moscow, Russia. He is part of the Aerospace Engineering Department and is responsible for the Electromagnetic Compatibility Laboratory. His research areas include electromagnetic compatibility, certification tests, space systems, and space technology development. Likewise, he is the technical responsible for the Space Science and Technology Theme Network from the National Council of Science and Technology from the Government of Mexico.

Rafael G. Chávez Moreno is an assistant professor at the National University Autonomous of Mexico School of Engineering. He received his Ph.D. in Mechanical Engineering from the School of Engineering-UNAM. He is part of the Aerospace Engineering Department. He is responsible for the Model-Based Design lab, which belongs to the Space and Automotive Engineering National Laboratory located at Juriquilla. He is an active member of the Mexican Society of Mechanical Engineering and the Space Science and Technology Network member. His current research areas include space, embedded, and control systems.

Jose Alberto Ramirez Aguilar is an assistant professor at the National Autonomous University of Mexico School of Engineering. He received his Ph.D. in Technical Sciences in Radio receivers and microsatellites from the Moscow Aviation Institute—M.A.I., Russian Federation. He is the head of the Aerospace Engineering Department. His current research areas are Radio Frequency and Microwave Systems, GNSS, Antennas, TT&C, Nano, and Microsatellites. Likewise, in 2020, he was selected for the first Latin American manned space mission, ESAA-01EX SOMINUS AD ASTRA.

Bryanda Guadalupe Reyes-Tesillo is a Chemical Industrial Engineer born in Mexico City. She received his Master of Materials Science in the line investigation of polymer materials from the Universidad Nacional Autónoma de Mexico. Her research is about hybrid coatings for aerospace applications.

Alfredo Rivas Iturbe is a Mechanical Engineer with Mechatronics training at "UNAM, Facultad de Estudios Superiores Aragón." Currently, with the advice of Dr. Jorge Alfredo Ferrer Pérez, he is in the Master's Degree in Mechatronics at Universidad Nacional Autónoma de México at his campus "UAT of Juriquilla," developing the topic "Analysis of the thermal behavior of a thermal vacuum chamber for compliance with space systems tests under ISO 19683: 2017 standard".

Bárbara Bermúdez-Reyes received a Ph. D. in Metallurgy and Materials Science in 2009. She is a Professor at the High Technology Unit of the National University Autonomous of Mexico. Previously, she was a Research Professor at different universities. She was General Secretary of the University Space Engineering Consortium-Mexico Chapter for 2014– 2019. She is a member of the Space Science and Technology Network. She has collaborated on projects with the Mexican Air Force and has evaluated projects of the Mexican Naval Research Institute. She is part of the National Research System with level I. Her research lines are designing and processing aerospace ceramic-ceramic and metal-ceramic composite materials, satellite structures, and severe plastic deformation to form aluminum alloy for high-velocity aircraft.

Dafne Gaviria Arcila is a professor at the National Autonomous University of Mexico. She has been involved in thermal and fluid analysis for over seven years. She was awarded 2020 her Ph.D. from the University of Nottingham in England. As part of her Ph.D. project, she received the ZONTA International Amelia Earhart Fellowship in 2017. Because of this recognition, in 2018, she received the Honorable Mention of Carlos Fuentes Award from the Mexican Embassy in the United Kingdom. She has participated in different activities to influence women to study STEM careers, such as mentoring, inspiring, and empowering them. Her story has been featured in the 100 Extraordinary Mexicans edition of the book titled Good Night Stories for Rebel Girls.

The manufacturer's authorised representative in the EU is Springer Nature Customer Service Centre GmbH, Europaplatz 3, 69115 Heidelberg, Germany. If you have any concerns regarding our products, please contact ProductSafety@springernature.com

Printed and bound by CPI Group (UK) Ltd, Croydon, CR0 4YY
05/10/2026
02228157-0005